COMPUTER ARITHMETIC ALGORITHMS

Second Edition

COMPUTER ARITHMETIC ALGORITHMS

Second Edition

Israel Koren
University of Massachusetts, Amherst

A K Peters
Natick, Massachusetts

Editorial, Sales, and Customer Service Office

A K Peters, Ltd.
63 South Avenue
Natick, MA 01760
www.akpeters.com

Library of Congress Cataloging-in-Publication data

Koren, Israel, 1945-
    Computer arithmetic algorithms / Israel Koren.–2nd ed.
        p. cm.
    Includes bibliographical references and index.
    ISBN 1-56881-160-8
     1. Computer arithmetic. 2. Computer algorithms. I. Title

  QA76.9.C62 K67 2001
  005.1–dc21

2001045837

Figures 4.3, 4.4, 4.5 and 4.7 are reprinted from D.J. Kuck, *The Structure of Computers and Computations*, Vol. 1 (copyright © 1978 John Wiley) by permission of John Wiley & Sons, Inc. Table 2.2 and Figures 5.32, 6.11, 6.13, 6.20–6.23, 7.5, 7.7, 7.13, and 10.2 are reprinted by permission of the Institute of Electrical and Electronics Engineers (IEEE). Table 4.3 is represented by permission of the IBM Systems Journal.

The first edition of this book was published by Prentice-Hall Inc.

Printed in Canada
06 05 04 03 02   10 9 8 7 6 5 4 3 2 1

This book is dedicated to my wife, Zahava,
my sons, Yuval and Yaron,
and to the memory of my parents,
Jacob and Dvora.

CONTENTS

vii

FORWORD TO THE
SECOND EDITION

This edition includes several new sections as well as many amendments and corrections made since the first edition in 1993. The new sections include floating-point adders, floating-point exceptions, general carry-look-ahead adders, prefix adders, Ling adders, and fused multiply-add units. New algorithms and implementations have been added to almost all chapters. My thanks to the students and readers whose discoveries of errors, and general comments, are reflected in this volume.

Since the first edition, a web page was created for *Computer Arithmetic Algorithms*, which contains updates on the book, solutions to selected problems and relevant links. It can be found at http://www.ecs.umass.edu/ece/koren/arith.

Additionally, there is now an on-line, JavaScript-based simulator for many of the algorithms contained in this book. A useful tool for both students and practitioners in the field, the Java-based simulator can be found at http://www.ecs.umass.edu/ece/koren/arith/simulator.

I very much welcome further comments and suggestions; please e-mail them to me at koren@ecs.umass.edu.

Israel Koren
Amherst, Massachusetts
July 2001

xi

PREFACE

The goal of this textbook is to explain the fundamental principles of algorithms available for performing arithmetic operations in digital computers. These include basic arithmetic operations like addition, subtraction, multiplication, and division in fixed-point and floating-point number systems, and more complex operations, such as square root extraction and evaluation of exponential, logarithmic, and trigonometric functions, and so on. Even the seemingly simple arithmetic operations turn out to be more complex than one expects when attempting to implement them. The descriptions found in many excellent books on computer architecture do not provide the level of detail required, and there is a need for a book that is completely devoted to digital computer arithmetic.

Designs that include arithmetic units have proliferated in recent years. The progress already made in integrated circuit technology, and further advances expected in the near future, has had a significant impact on the design of new arithmetic processors. The higher density of integrated circuits now enables the design and implementation of sophisticated arithmetic processors employing algorithms that were considered prohibitively complex in the past. Consequently, methods that were previously considered unconventional should be examined, since they may now be attractive alternatives. Furthermore, the ease of designing application-specific integrated circuits currently allows engineers to design arithmetic units tailored to their special needs, rather than having to use general-purpose circuits. These specialized arithmetic units can achieve a higher speed of operation for the particular application being considered.

This book describes the principles of computer arithmetic algorithms independently of any particular technology employed for their implementation. The existence of several implementation technologies and the rapid changes in these make a detailed description of any implementation almost immediately obsolete. The book includes numerical examples to illustrate the working of the algorithms presented and explains the concepts behind the algorithms without relying on gate diagrams. Such diagrams are usually straightforward, and any reader with a sufficiently good background in digital design, as is expected from the readers of this book, can draw his/her own. These diagrams are in many cases useless for the practitioner, since the technology that he/she plans to use imposes its constraints on the implementation, and, more importantly, they do not provide any additional insight.

All the algorithms in this book are described within the same framework, so that the similarities between different algorithms become evident, and, consequently, the basic principles behind these algorithms can be easily identified. This should help the reader to better understand the currently available algorithms, know how to select the most appropriate algorithm to match a given technology, and even be able to develop new algorithms if the need arises.

This book is intended to be used as a textbook in a senior-level or first-year graduate-level course in computer arithmetic and as a reference book for practicing engineers. The reader's expected background includes a basic knowledge of digital design and the principles of digital computer organization. The book includes 11 chapters; each chapter has a list of relevant references and a set of exercises. A separate solution manual is available from the publisher upon request.

This book does not include all the algorithms that have ever been suggested, and is meant only to serve as a solid introduction to this rich field, which is continuously evolving. Readers who are interested in further details on a particular topic should consult the list of references at the end of each chapter. Excellent sources for additional information are the proceedings of the biannual IEEE Symposium on Computer Arithmetic and periodicals such as the IEEE Journal on Solid-State Circuits or the IEEE Transactions on Computers. The latter had several special issues devoted to computer arithmetic. Other sources are the several books on computer arithmetic already available, which are listed at the end of Chapter 1.

This textbook evolved from lecture notes prepared for courses in computer arithmetic that I have taught at the University of California at Santa Barbara, the University of Southern California, the Technion in Israel, the University of California at Berkeley, and the University of Massachusetts at Amherst. The order of topics in the book follows their order in my lectures. However, several instructors who used a preliminary version of the book, did not experience any difficulties when covering the chapters out of order.

Many people have contributed in different ways to this book. Prof. James Howard from the University of California at Santa Barbara was the first to suggest the writing of this book. Prof. William Kahan from the University of California at Berkeley, with whom I had long discussions in 1983, influenced my views on many topics. Moshe Gavrielov from LSI Logic read much of the book, and his many suggestions helped to improve the presentation. Prof. Mary Jane Irwin from the Pennsylvania State University and Prof. Behrooz Parhami from the University of California at Santa Barbara agreed to use a preliminary version of this book in their class and provided many helpful comments and suggestions.

Several others reviewed parts of the book and gave me very important feedback. These include Prof. Dan Atkins from the University of Michigan at Ann Arbor, Prof. Milos Ercegovac from the University of California at Los Angeles, Prof. Earl Swartzlander, Tom Callaway, and Michael Schulte from the University of Texas at Austin, and Dr. George Taylor from Sun Microsystems.

The graduate students who took my course in the previously mentioned campuses made many contributions through their questions and suggestions. In particular, I wish to acknowledge the contributions made by Sachin Ghanekar, who prepared the solution manual, and by Ofra Zinaty, Moshe Gavrielov, and Susan Morin.

Last but not least, I wish to thank my wife, Zahava and my sons, Yuval and Yaron, who provided moral support as well as editorial assistance.

Israel Koren
Amherst, Massachusetts

1

CONVENTIONAL NUMBER SYSTEMS

1.1 THE BINARY NUMBER SYSTEM

In conventional digital computers, integers are represented as binary numbers of fixed length. A binary number of length n is an ordered sequence

$$(x_{n-1}, x_{n-2}, \cdots, x_1, x_0)$$

of binary digits where each digit x_i (also known as a *bit*) can assume one of the values 0 or 1. The length n of the sequence is of significance, since binary numbers in digital computers are stored in registers of a fixed length, n. The above sequence of n digits (or n-tuple) represents the integer value

$$X = x_{n-1}2^{n-1} + x_{n-2}2^{n-2} + \cdots + x_1 2 + x_0 = \sum_{i=0}^{n-1} x_i 2^i. \qquad (1.1)$$

Upper case letters are used in this book to represent numerical values or sequences of digits while lower case letters, usually indexed, represent individual digits. The weight of the digit x_i in (1.1) is the ith power of 2, which is called the *radix* of the number system. The interpretation rule in Equation (1.1) is similar to the rule used for the ordinary decimal numbers. There are, however, two differences between these interpretation rules. First, the radix 10 is used instead of 2 in Equation (1.1) and consequently, the allowed digits in the decimal case are $x_i \in \{0, 1, 2, \cdots, 9\}$ instead of $x_i \in \{0, 1\}$. We call the decimal numbers radix-10

1

numbers and the binary numbers radix-2 numbers. We indicate the radix to be used when interpreting a given sequence of digits by writing it as a subscript. Thus, the sequence $(101)_{10}$ represents the decimal value 101, while the sequence $(101)_2$ represents the decimal value 5.

Since operands and results in an arithmetic unit are stored in registers of a fixed length, there is a finite number of distinct values that can be represented within an arithmetic unit. Let X_{min} and X_{max} denote the smallest and largest representable values, respectively. We say that $[X_{min}, X_{max}]$ is the *range* of the representable numbers. Any arithmetic operation that attempts to produce a result larger then X_{max} (or smaller than X_{min}), will produce an incorrect result. In such cases the arithmetic unit should indicate that the generated result is in error. This is usually called an *overflow* indication.

Example 1.1

If the conventional binary number system is employed to represent unsigned integers using four binary digits (bits) then, $X_{max} = (15)_{10}$ is represented by $(1111)_2$ and $X_{min} = (0)_{10}$ is represented by $(0000)_2$. Increasing X_{max} by 1 results in $(16)_{10} = (10000)_2$ out of which, in a 4-bit representation, only the last four digits are retained, yielding $(0000)_2 = (0)_{10}$. In general, a number X that is not in the range $[X_{min}, X_{max}] = [0, 15]$ is represented by X modulus 16, or X mod 16, which is the remainder when dividing X by 16. Such a situation can arise, for example, when two operands X and Y are added and their sum exceeds X_{max}. In this case, the final result S satisfies $S = (X + Y)$ mod 16. For example,

$$
\begin{array}{r|ccccc}
X & 1 & 0 & 1 & 1 & 11 \\
+Y & 0 & 1 & 1 & 1 & 7 \\
\hline
 & 1\ 0 & 0 & 1 & 0 &
\end{array}
$$

Since the final result has to be stored in a 4-bit register, the most significant bit (whose weight is $2^4 = 16$) is discarded, yielding $(0010)_2 = 2$ = 18 mod 16.   □

1.2 MACHINE REPRESENTATIONS OF NUMBERS

The conventional binary system is a specific example of a number system that can be used to represent numerical values in an arithmetic unit. A number system is in general defined by the set of values that each digit can assume and by an interpretation rule that defines the mapping between the sequences of digits and their numerical values. We distinguish between conventional number systems like the binary system described in the previous section (or the commonly used decimal system) and unconventional systems like the signed-digit number system (to be presented in Chapter 2).

The conventional number systems are *nonredundant, weighted,* and *positional* number systems. In a *nonredundant* number system every number has a unique representation; in other words, no two sequences have the same numerical value. The term *weighted* number system means that there is a sequence of weights

$$w_{n-1}, w_{n-2}, \cdots, w_1, w_0$$

that determines the value of the given n-tuple $(x_{n-1}, x_{n-2}, \cdots, x_0)$ by the equation

$$X = \sum_{i=0}^{n-1} x_i w_i. \tag{1.2}$$

Thus, w_i is the weight assigned to the digit in the ith position, x_i. Finally, in a *positional* number system, the weight w_i depends only on the position i of the digit x_i. In the conventional number systems the weight w_i is the ith power of a fixed integer r, which is the *radix* of the number system. In other words, $w_i = r^i$. Therefore, these number systems are also called *fixed-radix* systems. Since the weight assigned to the digit x_i is r^i, this digit has to satisfy $0 \leq x_i \leq r - 1$. Otherwise, if $x_i \geq r$ is allowed, then

$$x_i r^i = (x_i - r) r^i + 1 \cdot r^{i+1},$$

resulting in two machine representations for the same value: $(\cdots, x_{i+1}, x_i, \cdots)$ and $(\cdots, x_{i+1} + 1, x_i - r, \cdots)$. In other words, allowing $x_i \geq r$ introduces redundancy into the fixed-radix number system.

A sequence of n digits in a register does not necessarily have to represent an integer. We may use such a sequence to represent a mixed number that has a fractional part as well as an integral part. This is done by partitioning the n digits into two sets: k digits in the integral part and m digits in the fractional part, satisfying $k + m = n$. The value of an n-tuple with a radix point between the k most significant digits and the m least significant digits

$$\big(\underbrace{x_{k-1} x_{k-2} \cdots x_1 x_0}_{integral\ part} \quad \cdot \quad \underbrace{x_{-1} x_{-2} \cdots x_{-m}}_{fractional\ part}\big)_r$$

is

$$\begin{aligned} X &= x_{k-1} r^{k-1} + x_{k-2} r^{k-2} + \cdots + x_1 r + x_0 + x_{-1} r^{-1} + \cdots + x_{-m} r^{-m} \\ &= \sum_{i=-m}^{k-1} x_i r^i. \end{aligned} \tag{1.3}$$

The radix point is not stored in the register but is understood to be in a fixed position between the k most significant digits and the m least significant digits. For this reason we call such representations *fixed-point* representations. A programmer of a digital computer is not necessarily restricted to the use of numbers having the predetermined position of the radix point but can properly scale the operands. As long as the same scaling factor is used for all operands, the add and subtract operations yield the correct results, since $aX \pm aY = a(X \pm Y)$, where a is the scaling factor. However, corrections are required when performing multiplication and division, since $aX \cdot aY = a^2 XY$ and $aX/aY = X/Y$. Commonly used positions for the radix point are at the rightmost side of the number (i.e., pure integers, $m = 0$) and at the leftmost side of the number (i.e., pure fractions, $k = 0$).

Given the length n of the operands, the weight r^{-m} of the least significant digit indicates the position of the radix point. To simplify our discussion from this point on and to avoid the need to distinguish between the different partitions of numbers (into fractional and integral parts), we introduce the notion of a *unit in the last position* (*ulp*), which is the weight of the least significant digit,

$$ulp = r^{-m}. \tag{1.4}$$

1.3 RADIX CONVERSIONS

Radix conversion is the translation of a number X represented in one radix number system (to be called the *source* number system) to its representation in another number system (called the *destination* number system). The major reason for such conversions is the fact that most arithmetic units operate on binary numbers while their users are more accustomed to decimal numbers, which also require a smaller number of digits. We will therefore emphasize conversions between the decimal and the binary number systems but will present the algorithms in a more general form.

Given a number X, we wish to find its representation in the destination number system with radix r_D. For convenience, we distinguish between the conversion of the integral part X_I and that of the fractional part X_F. Starting with the integral part, its representation $(x_{k-1} x_{k-2} \cdots x_1 x_0)_{r_D}$ is sought. We can rewrite the appropriate part of Equation (1.3) as follows:

$$X_I = \{[\cdots (x_{k-1} r_D + x_{k-2}) r_D + \cdots + x_2] r_D + x_1\} r_D + x_0, \tag{1.5}$$

where $0 \le x_i < r_D$. Thus, if we divide X_I by r_D we obtain x_0 as the remainder and

$$\{[\cdots (x_{k-1} r_D + x_{k-2}) r_D + \cdots + x_2] r_D + x_1\}$$

as the quotient. If we now divide the above quotient by r_D, we obtain x_1 as the remainder. We may therefore divide the quotients by r_D repeatedly, retaining the remainders as the required digits until a zero quotient is reached.

To find the representation $(x_{-1}x_{-2} \cdots x_{-m})_{r_D}$ of the fractional part X_F, we rewrite the appropriate part of Equation (1.3) as follows:

$$X_F = r_D^{-1} \left\{ x_{-1} + r_D^{-1} \left[x_{-2} + r_D^{-1}(x_{-3} + \cdots) \right] \right\}. \tag{1.6}$$

If we multiply X_F by r_D, we obtain a mixed number with x_{-1} as its integral part and

$$r_D^{-1} \left[x_{-2} + r_D^{-1}(x_{-3} + \cdots) \right]$$

as the fractional part. We may therefore multiply the fractional parts by r_D repeatedly, retaining the generated integers as the required digits. However, unlike the algorithm for the integral part, this algorithm is not guaranteed to terminate, since a finite fraction (one that needs a finite number of nonzero digits) in one number system may correspond to an infinite fraction in another. This does not constitute a problem in practice, since the process can be terminated after m steps (or a few additional ones if rounding is desired).

Example 1.2

The decimal mixed number $X = 46.375_{10}$ is to be converted to binary form. Starting with $X_I = 46$ we obtain the following quotients and remainders when repeatedly dividing by 2:

Quotient	Remainder
23	$0 = x_0$
11	$1 = x_1$
5	$1 = x_2$
2	$1 = x_3$
1	$0 = x_4$
0	$1 = x_5$

We now convert the fractional part $X_F = 0.375$ and obtain the following integers and fractions when repeatedly multiplying by 2:

Integer part	Fractional part
$0 = x_{-1}$	.75
$1 = x_{-2}$	.5
$1 = x_{-3}$	.0

Thus, the final result is $46.375_{10} = 101110.011_2$.

If the fractional part of the given decimal number was $X_F = 0.3$, the above algorithm would never terminate, since the decimal fraction 0.3_{10} is the infinite binary fraction $(0.0100110011 \cdots)_2$. $\square$

All the arithmetic operations in the above conversion algorithms were performed in the source number system, which was the decimal system. To convert a binary number to the decimal system, we may either execute the above algorithm in the source binary system or, more conveniently, perform the conversion in the destination decimal system using Equation (1.3).

1.4 REPRESENTATIONS OF NEGATIVE NUMBERS

For fixed-point numbers in a radix r system, we have to determine the way negative numbers are represented. Two different forms are commonly used:

1. Sign and magnitude representation, which is also called the signed-magnitude method

2. Complement representation which comprises two alternatives

1. Signed-magnitude: Here the sign and magnitude are represented separately. The first digit is the sign digit and the remaining $(n-1)$ digits represent the magnitude. In the binary case, the sign bit is normally selected to be 0 for positive numbers and 1 for negative ones. In the nonbinary case, the values 0 and $(r-1)$ are assigned to the sign digit of positive and negative numbers, respectively. Notice that in this case only $2 \cdot r^{n-1}$ out of the r^n possible sequences are utilized. This will be discussed further later on.

Let the $(n-1)$ digits representing the magnitude be partitioned into $(k-1)$ and m digits in the integral and fractional parts, respectively. The largest representable value is then

$$X_{max} = \left(r^{k-1} - ulp\right), \qquad \text{where} \quad ulp = r^{-m}$$

and the corresponding representation is $0(r-1)\cdots(r-1)$. Thus, the range of positive numbers is $\left[0, r^{k-1} - ulp\right]$. The range of negative numbers is similarly $\left[-(r^{k-1} - ulp), -0\right]$, represented by $(r-1)(r-1)\cdots(r-1)$ to $(r-1)0\cdots0$. We therefore have two representations for zero, one positive and one negative. This is inconvenient when implementing an arithmetic unit, since an *equal* indication must be generated in a test for zero operation for the two different representations of zero.

Example 1.3

In the binary case all 2^n sequences are utilized. The 2^{n-1} sequences from $00\cdots0$ to $01\cdots1$ represent positive numbers, while the remaining 2^{n-1} sequences from $10\cdots0$ to $11\cdots1$ represent negative numbers. If $k = n$ (and therefore, $m = 0$ and $ulp = 2^0 = 1$) the range of positive numbers is $\left[0, 2^{n-1} - 1\right]$ and the range of negative numbers is $\left[-(2^{n-1} - 1), -0\right]$. $\square$

A major disadvantage of the signed-magnitude representation is that the operation to be performed may depend on the signs of the operands. For example, when adding a positive number X and a negative number $-Y$ (i.e., Y is the absolute value of the second operand), we need to perform the calculation $X + (-Y)$. If $Y > X$, we should obtain as a final result $-(Y - X)$. We therefore need to first calculate $Y - X$, i.e., switch the order of the operands and perform subtraction rather than addition, and then attach the minus sign. This results in a sequence of decisions that have to be made, costing excess control logic and execution time. This is avoided in the complement representation methods.

2. Complement representations: There are two alternatives:

(i) Radix complement (also called two's complement in the binary system)
(ii) Diminished-radix complement (called one's complement in the binary system)

In both complement methods, a positive number is represented in the same way as in the signed-magnitude method, whereas a negative number, $-Y$, is represented by $(R - Y)$ where R is a constant whose value we will determine next. Such a representation satisfies the basic identity

$$-(-Y) = Y, \tag{1.7}$$

since the complement of $(R-Y)$ is $R-(R-Y) = Y$. One of the major advantages of a complement representation (regardless of the exact value of R) is that no decisions have to be made before executing an addition or subtraction. In the previous example, where a positive and a negative number are to be added, the second operand is represented by $(R - Y)$. Therefore, the addition to be performed is

$$X + (R - Y) = R - (Y - X).$$

If $Y > X$ then the negative result $-(Y - X)$ is already represented in the same complement form; i.e., as $R - (Y - X)$, and there is no need to make any special decisions like interchanging the order of the two operands. However, if $X > Y$, the correct result should be $(X - Y)$ while $X + (R - Y) = R + (X - Y)$. The additional term, R, must be discarded, and the value of R should be selected to simplify or even completely eliminate this correction step.

Another requirement on the value selected for R is that the calculation of the complement $(R - Y)$ of a given number Y be a simple operation that can be done at a high speed. Before deciding on the value of R we define the complement of a single digit x_i, denoted by $\bar{x}_i$, as

$$\bar{x}_i = (r - 1) - x_i. \tag{1.8}$$

We denote by $\overline{X}$ the n-tuple $(\bar{x}_{k-1}, \bar{x}_{k-2}, \cdots, \bar{x}_{-m})$ obtained after complementing every digit in the sequence corresponding to X. We now add $\overline{X}$ to X and, based on Equation (1.8), we obtain $x_i + \bar{x}_i = (r - 1)$, independent of the exact value of x_i. We then add *ulp* to the sum of X and $\overline{X}$, yielding

X	x_{k-1}	x_{k-2}	$\cdots$	x_{-m}	
$+\overline{X}$	$\bar{x}_{k-1}$	$\bar{x}_{k-2}$	$\cdots$	$\bar{x}_{-m}$	
	$(r-1)$	$(r-1)$	$\cdots$	$(r-1)$	
$+ulp$				1	
1	0	0	$\cdots$	0	$= r^k$

The above calculation can be rewritten as

$$X + \overline{X} + ulp = r^k. \tag{1.9}$$

Note that when the above result is stored into a register of length n $(n = k+m)$, the most significant digit is discarded and the final result is zero. In general, storing the result of any arithmetic operation into a fixed-length register is equivalent to taking the remainder after dividing by r^k.

Rearranging the terms in the previous equation results in

$$r^k - X = \overline{X} + ulp.$$

Consequently, if we select the value r^k for R we obtain

$$R - X = r^k - X = \overline{X} + ulp. \tag{1.10}$$

The calculation of the complement $(R - X)$ of a given number X as defined above is quite simple and is independent of the value of k. We call this *radix-complement* representation. No correction is needed for it when the result of the previous operation, $X + (R - Y)$, is positive (i.e., when $X > Y$), since $R = r^k$ is discarded when calculating $R + (X - Y)$.

Example 1.4

For $r = 2$ and $k = n = 4$ (and consequently, $m = 0$ and $ulp = 2^0 = 1$) the radix complement (also called the two's complement in the binary case) of a number X equals $2^4 - X$ but can instead be calculated, according to Equation (1.10), by $\overline{X} + 1$. In this case, the sequences 0000 to 0111 represent the positive numbers 0_{10} to 7_{10}, respectively. The two's complement of the largest positive number is $1000+1 = 1001$ and it represents the value $(-7)_{10}$. The two's complement of zero is $1111 + 1 = 10000 = 0 \bmod 2^4$; i.e., there is a single representation of zero. Thus, each positive number has a corresponding negative number that starts with a 1. There is an

Sequence	Two's complement	One's complement	Signed-magnitude
0111	7	7	7
0110	6	6	6
0101	5	5	5
0100	4	4	4
0011	3	3	3
0010	2	2	2
0001	1	1	1
0000	0	0	0
1111	−1	−0	−7
1110	−2	−1	−6
1101	−3	−2	−5
1100	−4	−3	−4
1011	−5	−4	−3
1010	−6	−5	−2
1001	−7	−6	−1
1000	−8	−7	−0

TABLE 1.1   Three representation methods of binary numbers with $k = n = 4$.

additional sequence that starts with a 1, namely, 1000, which has no corresponding positive number. It represents the negative number $(-8)_{10}$. Therefore, the range of binary numbers in the two's complement method with $k = n = 4$ is $-8 \leq X \leq 7$. The two's complement representations of all values within this range are shown in Table 1.1.

To illustrate the simplicity of executing the operation $X + (-Y)$ with $Y > X$, consider the addition of the numbers 2 and −7, represented by 0010 and 1001, respectively:

$$
\begin{array}{ccccc}
  & 0 & 0 & 1 & 0 & \quad 2 \\
+ & 1 & 0 & 0 & 1 & \quad -7 \\
\hline
1 & 0 & 1 & 1 & & \quad -5
\end{array}
$$

This is the correct result represented in the two's complement method, and there is no need for any preliminary decisions or post corrections. Even when $X > Y$, the expected result is calculated without requiring any corrections. For example, when adding 7 and −2, represented by 0111 and 1110, respectively, we obtain

$$
\begin{array}{ccccc}
  & 0 & 1 & 1 & 1 & \quad 7 \\
+ & 1 & 1 & 1 & 0 & \quad -2 \\
\hline
1 & 0 & 1 & 0 & 1 & \quad 5
\end{array}
$$

Only the last four least significant digits are retained, yielding 0101.   □

A second possible choice for R is $R = r^k - ulp$. This is the *diminished-radix complement*, for which, according to Equation (1.9),

$$R - X = (r^k - ulp) - X = \overline{X}. \tag{1.11}$$

Here, the derivation of the complement is even simpler than that of the radix complement. All the digit-complements $\bar{x}_i$ can be calculated in parallel, leading to a fast computation of $\overline{X}$. On the other hand, a correction step is needed when the result $R + (X - Y)$ is obtained and $(X - Y)$ is positive, as will be explained later.

Example 1.5

For $r = 2$ and $k = n = 4$, the diminished-radix complement (also called the one's complement in the binary case) of a number X equals $(2^4 - 1) - X$, which also equals $\overline{X}$, the sequence of digit complements, according to Equation (1.11). As in the previous example, the sequences 0000 to 0111 represent the positive numbers 0 to 7, respectively. The one's complement of the largest positive number is 1000, representing the value $(-7)_{10}$. The one's complement of zero is 1111; i.e., there are two representations of zero. In summary, the range of binary numbers in the one's complement method with $k = n = 4$ is $-7 \le X \le 7$. The different representations of positive and negative numbers with $k = n = 4$ in the three methods are compared in Table 1.1.   □

In the binary case, the most significant digit can assume only two values and is thus a "true" sign digit. This holds for all three representation methods as shown in Table 1.1 and the distinction between positive and negative numbers is greatly simplified. In the nonbinary case, restricting the most significant digit to two values only (0 and $(r - 1)$) would considerably reduce the percentage of utilized sequences; only $2 \cdot r^{n-1}$ out of r^n (or 2 out of r) would be used. To make up for this, we can let the most significant digit assume all its possible values and partition the total number, r^n, of sequences equally (or almost equally) between positive and negative values. In general, a given number X is represented in a complement system by X if it is positive or by $R - |X|$ if it is negative. To have unambiguous representations, the regions for positive and negative numbers should not overlap. In other words, the inequality

$$|X| \le R/2$$

must be satisfied. If the value $X = R/2 + 1$ is allowed to be included in the region of representable numbers, then the negative number $-X$ is represented by $R - X = R/2 - 1$, which is identical to the representation of the positive number $R/2 - 1$. Nonoverlapping regions for positive and negative numbers can be achieved easily if the radix r is an even number. In this case, in order to satisfy the inequality $|X| \leq R/2 = \frac{r}{2} \cdot r^{n-1}$ (for an integer-only representation), the values $0, 1, \cdots, \frac{r}{2} - 1$ for the most significant digit would correspond to positive numbers, while the values $\frac{r}{2}, \cdots, r - 1$ would correspond to negative numbers. If, however, the radix is odd, then the representations 0 to $(\frac{r^n}{2} - 1)$ must be made positive and the remaining ones negative, making it more difficult to distinguish between positive and negative numbers.

Example 1.6

In the radix-complement decimal system the most significant digit can assume any of its 10 possible values. Thus, all sequences with a leading digit of 0, 1, 2, 3, or 4 represent positive numbers, while those having a leading digit of 5, 6, 7, 8, or 9 represent negative ones. For $n = 4$, the largest positive number is 4999 and the sequences 5000 through 9999 represent negative numbers with values of -5000 through -1, respectively. The range is therefore $-5000 \leq X \leq 4999$ and is shown in the diagram below.

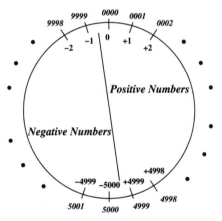

Let $Y = 2345$; to find the representation of $-Y = -2345$ (i.e., the radix complement $R - Y$ with $R = 10^4$) we use the expression $R - Y = \overline{Y} + ulp$. The digit complement in this case is the nine's complement, yielding $\overline{Y} = 7654$ and thus, $\overline{Y} + 1 = 7655$. We can verify this result by adding Y to $-Y$, obtaining $2345 + 7655 = 10^4 = 0 \bmod 10^4$.   □

The reader should realize by now that algorithms for arithmetic operations can be developed for various fixed-radix number systems and for different partitions of mixed numbers (into their integral and fractional parts). To simplify our discussion from this point on, we restrict ourselves to binary integers; i.e., $r = 2$ and $k = n$. The extension of what follows to the general case of binary numbers with $m \neq 0$ or radix r numbers with $r \neq 2$ is, in most cases, straightforward.

1.4.1 The Two's Complement Representation

The range of numbers in the two's complement system for $k = n$ is

$$- 2^{n-1} \leq X \leq (2^{n-1} - ulp) \quad \text{with} \quad ulp = 2^0 = 1.$$

This range is slightly asymmetric as there is one more negative number than there are positive numbers. The binary number $- 2^{n-1}$ (represented by $10 \cdots 0$) does not have a positive equivalent. Consequently, if a complement operation for this number is attempted, an overflow indication must be generated. On the other hand, there is a unique representation for 0, as shown in Table 1.1.

   Given a representation $(x_{n-1}, x_{n-2}, \cdots, x_0)$ in two's complement, we can use the following procedure to find its numerical value X: If $x_{n-1} = 0$, then $X = \sum_{i=0}^{n-1} x_i 2^i$, while if $x_{n-1} = 1$, the given sequence represents a negative number whose absolute value can be obtained by complementing the given sequence and then employing the previous equation.

Example 1.7

   Given the 4-tuple 1011, we can find its value by first complementing it to obtain $0100 + 1 = 0101$, then calculating the value of the sequence 0101, which is 5. This indicates that the value of the original sequence is -5. $\square$

   Instead of the above procedure we can use the expression

$$X = -x_{n-1} \cdot 2^{n-1} + \sum_{i=0}^{n-2} x_i 2^i. \tag{1.12}$$

Using Equation (1.12) for 1011 we obtain $-8 + 2 + 1 = -5$.

   To prove the validity of Equation (1.12) note first that, if $x_{n-1} = 0$, the exact same positive value is calculated. If $x_{n-1} = 1$, the value of the given representation is

$$- \left[\overline{X} + ulp \right] = - \left[\sum_{i=0}^{n-2} \bar{x}_i 2^i + 1 \right] = - \left[\sum_{i=0}^{n-2} (1 - x_i) 2^i + 1 \right]$$

$$= - \left[\sum_{i=0}^{n-2} 2^i - \sum_{i=0}^{n-2} x_i 2^i + 1 \right] = - \left[(2^{n-1} - 1) - \sum_{i=0}^{n-2} x_i 2^i + 1 \right]$$

$$= - 2^{n-1} + \sum_{i=0}^{n-2} x_i 2^i$$

which is exactly the value of the right-hand side of Equation (1.12) for $x_{n-1} = 1$.

1.4.2 The One's Complement Representation

The range of representable numbers in the one's complement system for $k = n$ is symmetric, and equals

$$-(2^{n-1} - ulp) \leq X \leq (2^{n-1} - ulp) \quad \text{with} \quad ulp = 2^0 = 1$$

As a result, there are two representations of zero, a positive zero represented by $000\cdots 0$, and a negative zero represented by $111\cdots 1$, as shown in Table 1.1.

For the one's complement system we have an equation similar to that of the two's complement system:

$$X = -x_{n-1}(2^{n-1} - ulp) + \sum_{i=0}^{n-2} x_i 2^i. \tag{1.13}$$

For example, the 4-tuple 1010 represents the value $-(2^3 - 1) + 2 = -5$. The proof of Equation (1.13) is left to the reader as an exercise.

The derivation of the one's complement is simpler than that of the two's complement. For each digit we have to calculate $\bar{x}_i = 1 - x_i$, which is the Boolean complement and can be done in parallel for all digits.

1.5 ADDITION AND SUBTRACTION

When adding or subtracting numbers represented in the signed-magnitude representation, only the magnitude bits participate in the arithmetic operation, while the sign bits are treated separately. Consequently, a carry-out (or borrow-out) indicates overflow. For example,

0		1	0	1	1	11	
0	+	0	1	1	0	6	
0		1	0	0	0	1	1 Carry-out

The final result is positive (the sum of two positive numbers), but its magnitude in four bits is erroneously obtained as 1 instead of 17, since $1 = 17 \bmod 16$.

In both complement systems, all digits, including the sign digit, participate in the add or subtract operation. A carry-out is therefore not necessarily an indication of an overflow in these systems. For example, when adding the two numbers $X = 13$ and $Y = -8$ represented in the two's complement method, we obtain

	0	1	1	0	1	13	
+	1	1	0	0	0	-8	
1	0	0	1	0	1	5	Carry-out but no overflow

The carry-out is discarded and does not indicate overflow. In general, if X and Y have opposite signs, no overflow can occur regardless whether there is a carry-out or not, as illustrated in the following two examples:

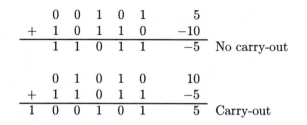

$$
\begin{array}{rcccccr}
  & 0 & 0 & 1 & 0 & 1 & \quad 5 \\
+ & 1 & 0 & 1 & 1 & 0 & -10 \\
\hline
  & 1 & 1 & 0 & 1 & 1 & -5 \quad \text{No carry-out}
\end{array}
$$

$$
\begin{array}{rcccccr}
  & 0 & 1 & 0 & 1 & 0 & 10 \\
+ & 1 & 1 & 0 & 1 & 1 & -5 \\
\hline
1 & 0 & 0 & 1 & 0 & 1 & 5 \quad \text{Carry-out}
\end{array}
$$

If X and Y have the same sign and the sign of the result is different from that of the two operands, then an overflow occurs. For example,

$$
\begin{array}{rcccccr}
  & 1 & 1 & 0 & 0 & 1 & -7 \\
+ & 1 & 0 & 1 & 1 & 0 & -10 \\
\hline
1 & 0 & 1 & 1 & 1 & 1 & 15 \quad \text{Carry-out and overflow}
\end{array}
$$

$$
\begin{array}{rcccccr}
  & 0 & 0 & 1 & 1 & 1 & 7 \\
+ & 0 & 1 & 0 & 1 & 0 & 10 \\
\hline
  & 1 & 0 & 0 & 0 & 1 & -15 \quad \text{No carry-out but overflow}
\end{array}
$$

In the one's complement system a carry-out is an indication that a correction step is needed. For example, when adding a positive number X and a negative number $-Y$ represented in one's complement, the result is

$$
X + (2^n - ulp) - Y = (2^n - ulp) + (X - Y)
$$

and if $X > Y$ we should obtain $(X - Y)$. The term 2^n represents the carry-out bit, which is discarded since the final result should be stored into a register of length n. The result is therefore $(X - Y - ulp)$, and the necessary correction is made by adding ulp. For example,

$$
\begin{array}{rcccccr}
                    & 0 & 1 & 0 & 1 & 0 & 10 \\
+   & 1 & 1 & 0 & 1 & 0 & -5 \\
\hline
1 \quad   & 0 & 0 & 1 & 0 & 0 & \\
\text{Correction} + &   &   &   &   & 1 & ulp \\
\hline
                    & 0 & 0 & 1 & 0 & 1 & 5
\end{array}
$$

The generated carry-out is called *end-around carry*, meaning that the carry-out is an *indication* that a 1 should be added to the least significant position. If there is no carry-out then no correction is needed. This is the case when $X < Y$ and thus the result, $-(Y - X)$, is negative and should be represented by $(2^n - ulp) - (Y - X)$. For example,

$$
\begin{array}{rcccccr}
  & 0 & 0 & 1 & 0 & 1 & 5 \\
+ & 1 & 0 & 1 & 0 & 1 & -10 \\
\hline
  & 1 & 1 & 0 & 1 & 0 & -5 \quad \text{No carry-out and hence no correction}
\end{array}
$$

As we have seen before, no such correction is necessary in two's complement addition.

In both complement systems, a subtract operation, $X - Y$, is performed by adding the complement of Y to X. In the one's complement system this means adding $\overline{Y}$ to X; in other words, $X - Y = X + \overline{Y}$. In the two's complement system we perform $X - Y = X + (\overline{Y} + ulp)$. This still requires only a single adder operation, since ulp is added through the forced carry input to the binary adder. This will be further explained in Chapter 5.

1.6 ARITHMETIC SHIFT OPERATIONS

Another way of distinguishing among the three methods for representing negative numbers is to consider the infinite extensions to the right and to the left of a given number. In the signed-magnitude method, the magnitude $x_{n-2}, \cdots, x_0$ can be viewed as the infinite sequence

$$\cdots 0, 0, \{x_{n-2}, \cdots, x_0\}, 0, 0 \cdots$$

If any arithmetic operation results in a nonzero prefix, then this constitutes an overflow.

In the radix-complement scheme the infinite extension is

$$\cdots x_{n-1}, x_{n-1}, \{x_{n-1}, \cdots, x_0\}, 0, 0 \cdots$$

where x_{n-1} is the sign digit. Finally, for the diminished-radix complement scheme the sequence is

$$\cdots x_{n-1}, x_{n-1}, \{x_{n-1}, \cdots, x_0\}, x_{n-1}, x_{n-1} \cdots$$

Example 1.8

The sequences 1011., 11011.0, and 111011.00 all represent the value -5 in the two's complement method. Similarly, the sequences 1010., 11010.1, and 1110101.11 all represent the value -5 in the one's complement method. The general proof is left as an exercise for the reader.   □

These extensions are useful when adding operands with different numbers of bits. The shorter operand must be extended to the length of the longer one before being added. Based on these extensions we can derive the rules for arithmetic shift operations where a left and right shift are equivalent to multiplication and division by 2, respectively.

Example 1.9

In two's complement,

$$Sh.L \ \{00101_2 = +5\} \ = \ 01010_2 \ = \ +10$$
$$Sh.R \ \{00101_2 = +5\} \ = \ 00010_2 \ = \ +2$$
$$Sh.L \ \{11011_2 = -5\} \ = \ 10110_2 \ = \ -10$$
$$Sh.R \ \{11011_2 = -5\} \ = \ 11101_2 \ = \ -3$$

In one's complement,

$$Sh.L \ \{11010_2 = -5\} \ = \ 10101_2 \ = \ -10$$
$$Sh.R \ \{11010_2 = -5\} \ = \ 11101_2 \ = \ -2$$

$\Box$

Arithmetic shift operations are very useful in many algorithms for multiplication and division, as will be described in Chapter 3.

1.7 EXERCISES

1.1. (a) Find the fixed-point representations of the two values $(41.25)_{10}$ and $(-41.25)_{10}$ in the radix complement and diminished-radix complement systems if the radix r is 8, $k = 3$ integer digits, and $m = 1$ fraction digit.
(b) Repeat (a) for $r = 2$, $k = 8$ and $m = 2$.

1.2. Verify the correctness of the following procedure to obtain the two's complement, $(y_{n-1}, y_{n-2}, \cdots, y_0)$ of a given sequence $(x_{n-1}, x_{n-2}, \cdots, x_0)$.
Start from the rightmost bit x_0. For each 0 in X set the corresponding bit in Y to 0 until you reach the first 1. For this 1 set the corresponding bit to 1. From this point on set $y_i = \bar{x}_i$.

1.3. Show that approximately $3.3n$ bits are needed to represent an n-digit decimal number.

1.4. In this problem we attempt to find the most "efficient" fixed-radix number system. Assume that N different values need to be represented and search for the radix r that minimizes the product $E = n \cdot r$ where n is the number of radix r digits that are required to represent N values; i.e., $n = \log_r N$. Thus, the function $E = r \cdot \log_r N = (r/\log_{10} r) \cdot \log_{10} N$ must be minimized.
(a) Justify the selection of the objective function E.
(b) Tabulate the values of $r/\log_{10} r$ for $r = 2, e, 3, \cdots, 10$. What are the best choices for r?

1.5. Prove the extensions of two's complement and one's complement numbers to infinite sequences. In particular, based on Equation (1.12) show that the representation of a signed binary number in $(n + 1)$ bits in the two's complement method may be derived from its representation in n bits by repeating the sign bit.

1.6. Given an n-tuple $X = (x_{n-1}, \cdots, x_0)$ find the value of $Sh.R\{X\} + Sh.R\{-X\}$ and of $Sh.L\{X\} + Sh.L\{-X\}$ using either the diminished-radix complement or the radix complement representation.

1.7. What are the rules for overflow detection in add/subtract operations when using two's complement or one's complement representations, assuming that carry-in and carry-out signals for the sign digit are available?

1.8. Prove that the value of any number in the one's complement system can be calculated, for any k and m, using the formula

$$X = -x_{k-1}(2^{k-1} - ulp) + \sum_{i=-m}^{k-2} x_i 2^i$$

1.9. Can the equation $X = -x_{n-1}2^{n-1} + \sum_{i=0}^{n-2} x_i 2^i$ be extended to the radix complement representation for any positive radix r?

1.10. The difference $X - Y$ can be formed by adding the complement of X to Y and then complementing the result. Prove that this is true for any complement scheme and any general fixed-radix number system. Is it useful?

1.11. Show that a carry-out in any add or subtract operation of one's complement numbers indicates that an end-around carry must be added to correct the final result. Pay particular attention to the case where two negative numbers need to be added.

1.12. Can the correction step in one's complement addition by adding an end-around carry generate another end-around carry?

1.13. In the signed-magnitude representation there is another way to perform the operation $X + (-Y)$ when $Y > X$. Instead of changing the order of operands and calculating $-(Y - X)$, we can simply subtract Y from X. However, this will result in a need for a correction. For example,

0		1	0	1	1	+11
1	−	1	1	0	1	−13
1		1	1	1	0	−14

while the correct result is 1 0010 $= (-2)_{10}$. Show that the required correction can be done by taking the two's complement of the result 1110. Explain why the complement operation provides the necessary correction.

1.8 REFERENCES

A number of textbooks focusing on computer arithmetic have been published since 1963. These include:

[1] J. J. F. CAVANAGH, *Digital computer arithmetic: Design and implementation,* McGraw-Hill, New York, 1984.

[2] I. FLORES, *The logic of computer arithmetic,* Prentice Hall, Englewood Cliffs, NJ, 1963.

[3] M. J. FLYNN AND S. F. OBERMAN, *Advanced computer arithmetic design,* Wiley, New York, 2001.

[4] J. B. GOSLING, *Design of arithmetic units for digital computers,* Springer-Verlag, New York, 1980.

[5] K. HWANG, *Computer arithmetic: Principles, architecture, and design,* Wiley, New York, 1978.

[6] J-M. MULLER, *Elementary Functions: Algorithms and Implementations,* Birkhäuser, Boston, 1997.

[7] A. R. OMONDI, *Computer arithmetic systems: Algorithms, architecture and implementations,* Prentice Hall, Englewood Cliffs, NJ, 1994.

[8] B. PARHAMI, *Computer arithmetic: Algorithms and Hardware Designs,* Oxford University Press, New York, 2000.

[9] N. R. SCOTT, *Computer number systems and arithmetic,* Prentice Hall, Englewood Cliffs, NJ, 1985.

[10] O. SPANIOL, *Computer arithmetic: Logic and design,* Wiley, New York, 1981.

[11] S. WASER and M. J. FLYNN, *Introduction to arithmetic for digital system designers,* Holt, Rinehart, Winston, New York, 1982.

Reprints of many classic papers appear in the next two volumes:

[12] E. E. SWARTZLANDER, JR. (Editor), *Computer arithmetic,* vol. 1, IEEE Computer Society Press, Los Alamitos, CA, 1990.

[13] E. E. SWARTZLANDER, JR. (Editor), *Computer arithmetic,* vol. 2, IEEE Computer Society Press, Los Alamitos, CA, 1990.

Several chapters in computer organization textbooks and some survey articles have been devoted to computer arithmetic including:

[14] Y. CHU, *Computer organization and microprogramming,* Prentice Hall, Englewood Cliffs, NJ, 1972, chap. 5.

[15] H. L. GARNER, "Number systems and arithmetic," in *Advances in computers,* vol. 6, F. L. Alt and M. Rubinoff (Eds.), Academic, New York, 1965, pp. 131-194.

[16] H. L. GARNER, "Theory of computer addition and overflows," *IEEE Trans. on Computers, C-27* (April 1978), 297-301.

[17] V. C. HAMACHER, Z. G. VRANESIC, and S. G. ZAKY, *Computer organization,* 2nd ed., McGraw-Hill, New York, 1984.

[18] D. GOLDBERG, "Computer arithmetic," in *Computer architecture: A quantitative approach,* D. A. Patterson and J. L. Hennessy, 2nd edition, Morgan Kaufmann, San Mateo, CA, 1996.

[19] U. KULISCH, "Mathematical foundations of computer arithmetic," *IEEE Trans. on Computers, C-26* (July 1977), 610-620.

[20] O. L. MACSORLEY, "High-speed arithmetic in binary computers," *Proc. of IRE, 49* (Jan. 1961), 67-91.

[21] G. W. REITWIESNER, "Binary arithmetic," in *Advances in computers,* vol. 1, F. L. Alt (Editor), Academic, New York, 1960, pp. 231-308.

[22] C. TUNG, "Arithmetic," in *Computer science,* A. F. Cardenas et al. (Eds.), Wiley-Interscience, New York, 1972.

2

UNCONVENTIONAL FIXED-
RADIX NUMBER SYSTEMS

Although the conventional binary number system with two's complement representation of negative numbers is commonly used in arithmetic units, there are several other number systems which have proven to be useful for certain applications. These include the negative radix and signed-digit number systems, both of which are described in this chapter. Other unconventional number systems, including the sign-logarithm number system and the residue number system, are discussed in Chapters 10 and 11, respectively.

2.1 NEGATIVE-RADIX NUMBER SYSTEMS

Conventional number systems are fixed-radix systems for which the weight w_i of the ith digit, x_i, is r^i, the range of each digit is $\{0, 1, \cdots, r-1\}$ and the interpretation rule for calculating the numerical value of the sequence $(x_{n-1}, x_{n-2}, \cdots, x_0)$ is

$$X = \sum_{i=0}^{n-1} x_i \, w_i = \sum_{i=0}^{n-1} x_i \, r^i. \tag{2.1}$$

The radix r is normally selected to be a positive integer. However, it is not necessary to restrict r to positive values, and we may select $r = -\beta$, where β is a positive integer. The digit set remains the same; i.e., $x_i \in \{0, 1, \cdots, \beta - 1\}$. The value of the n-tuple $(x_{n-1}, x_{n-2}, \cdots, x_0)$ in this *negative-radix* system is

$$X = \sum_{i=0}^{n-1} x_i \, (-\beta)^i. \tag{2.2}$$

In other words, the weight w_i satisfies

$$w_i = \begin{cases} \beta^i & \text{if } i \text{ is even} \\ -\beta^i & \text{if } i \text{ is odd.} \end{cases}$$

Example 2.1

The negative-radix number system with $\beta = 10$ is called the *nega-decimal* system. Consider the following three-digit nega-decimal numbers:

$$(192)_{-10} = 100 - 90 + 2 = 12, \qquad (012)_{-10} = -10 + 2 = -8$$

The largest positive value that can be represented as a 3-tuple $(x_2, x_1, x_0)_{-10}$ is $(909)_{-10} = 909_{10}$ while the smallest is $(090)_{-10} = -90_{10}$. Thus, the range of values represented as 3-tuples (x_2, x_1, x_0) in the nega-decimal system is $-90 \leq X \leq 909$. This range is asymmetric, since there are approximately 10 times as many positive numbers as negative ones. This is always true for odd values of n. If n is even then the opposite is true. For example, the range for $n = 4$ is $-9090 \leq X \leq 909$.   □

In the negative-radix number system there is no need for a separate sign digit, and consequently there is no need for a special method like the radix-complement method, to represent negative numbers. The sign of the number is determined by the first nonzero digit. The fact that there is no distinction between positive number and negative number representations makes the arithmetic operations indifferent to the sign of the number. However, the algorithms for the basic arithmetic operations in the negative-radix number system are slightly more complex then their counterparts for the conventional number systems, as illustrated in the following example.

Example 2.2

Consider negative-radix numbers of length $n = 4$ with $\beta = 2$. This number system is called the *nega-binary* system. The range for 4-bit nega-binary numbers is

$$(-10)_{10} = (1010)_{-2} \leq X \leq (0101)_{-2} = (+5)_{10}.$$

When adding nega-binary numbers the carry bits can be either positive or negative as illustrated in the following addition, where the weights of the different bit positions are shown in the top row:

	-8	$+4$	-2	$+1$	
	0	1	0	1	5
+	1	1	0	1	-3
	0	1	1	0	2

Note that in the -2 column there is a carry-in whose weight is $+2$, and the only way to handle it is to convert it into $+4 - 2$; i.e., produce a sum bit whose weight is -2 and a positive carry-out bit to the $+4$ column. Also note that in the $+4$ column, a carry-out is generated with weight $+8$. This carry bit and the operand bit in the -8 column cancel each other out to produce a zero sum bit.   □

The nega-binary number system has been proposed for several signal processing applications, and algorithms for all arithmetic operations in this number system have been devised. However, its use has been limited and there are currently no standard integrated circuits (ICs) that perform arithmetic operations in this system. One of the reasons for this is the fact that it is not superior, in principle, to the more well established two's complement system. As shown in the next section, these two are members of a larger group of number systems with very similar properties.

2.2 A GENERAL CLASS OF FIXED-RADIX NUMBER SYSTEMS

The negative-radix, and many other fixed-radix number systems, are members of a broad class of nonredundant number systems. In this class, each n-digit number system is characterized by a positive radix β and a vector Λ of length n, $\Lambda = (\lambda_{n-1}, \lambda_{n-2}, \cdots, \lambda_0)$ where $\lambda_i \in \{-1, 1\}$. Such a system, with a standard digit set, $\{0, 1, \cdots, \beta - 1\}$, can be identified by the triplet $< n, \beta, \Lambda >$. The value X of an n-tuple $(x_{n-1}, x_{n-2}, \cdots, x_0)$ in the system $< n, \beta, \Lambda >$ is given by

$$X = \sum_{i=0}^{n-1} \lambda_i x_i \beta^i. \tag{2.3}$$

The multiplying factor λ_i allows us to select between the two possible weights β^i and $-\beta^i$, individually, for every digit position i.

For any given radix β there are in this class 2^n distinct number systems corresponding to the different values of Λ. Among them is the positive-radix number system, for which $\lambda_i = +1$ for every i, and the negative-radix number system, for which $\lambda_i = (-1)^i$ for every i. Also included is the radix-complement number system with x_{n-1} as a "true" sign digit (i.e., $x_{n-1} \in \{0, 1\}$) and the characterizing vector $\Lambda = (-1, 1, 1, \cdots, 1)$ [3].

We will now examine some of the properties of this general class of number systems. Let P and N denote the largest and smallest representable integers, respectively, in the general system $< n, \beta, \Lambda >$. The digits of $P = (p_{n-1}, p_{n-2}, \cdots p_0)$ satisfy

$$p_i = \begin{cases} \beta - 1 & \text{if } \lambda_i = +1 \\ 0 & \text{otherwise} \end{cases} \quad ; \quad i = 0, 1, \cdots, n-1$$

A more convenient expression for p_i is $p_i = \frac{1}{2}(\lambda_i + 1)(\beta - 1)$, based on which the value of the n-tuple $(p_{n-1}, p_{n-2}, \cdots p_0)$ is

$$\begin{aligned} P & = \sum_{i=0}^{n-1} \frac{1}{2}(\lambda_i + 1)(\beta - 1)\beta^i = \frac{1}{2}\left[\sum_{i=0}^{n-1} \lambda_i(\beta - 1)\beta^i + \sum_{i=0}^{n-1}(\beta - 1)\beta^i\right] \\ & = \frac{1}{2}[Q + (\beta^n - 1)] \end{aligned} \tag{2.4}$$

where Q is the value of the n-tuple $(\beta - 1, \beta - 1, \cdots, \beta - 1)$ in $< n, \beta, \Lambda >$. As will become evident later on, Q is a very significant quantity. Similarly, the digits of $N = (y_{n-1}, y_{n-2}, \cdots y_0)$, the smallest representable number, satisfy

$$y_i = \begin{cases} \beta - 1 & \text{if } \lambda_i = -1 \\ 0 & \text{otherwise} \end{cases} \quad ; \quad i = 0, 1, \cdots, n-1$$

and its value is

$$N = \sum_{i=0}^{n-1} \frac{1}{2}(\lambda_i - 1)(\beta - 1)\beta^i = \frac{1}{2}[Q - (\beta^n - 1)]. \tag{2.5}$$

The number of integers in the range $N \leq X \leq P$ is $P - N + 1 = \beta^n$, and the range is, in general, asymmetric. A measure of the asymmetry can be the difference between the absolute values of the largest and smallest numbers:

$$P - |N| = P + N = Q \tag{2.6}$$

Example 2.3

The negative-radix system for which $\Lambda = (\cdots, -1, +1, -1, +1)$ has an asymmetric range, and for an even value of n there are β times as many negative numbers as positive ones. If we prefer to have more positive numbers, we can instead use the system $< n, \beta, \Lambda = (\cdots, +1, -1, +1, -1) >$. Two of the binary systems are nearly symmetric: the two's complement system, $< n, \beta = 2, \Lambda = (-1, 1, 1, \cdots, 1) >$, for which $P + N = Q = -ulp$, and the system $< n, \beta = 2, \Lambda = (+1, -1, -1, \cdots, -1) >$ for which $P + N = Q = +ulp$. $\qquad \square$

The complement $\overline{X}$ of a number X in a system $< n, \beta, \Lambda >$ is defined based on the digit complement of x_i which is $\bar{x}_i = (\beta - 1) - x_i$, as follows:

$$\overline{X} = \sum_{i=0}^{n-1} \bar{x}_i \lambda_i \beta^i = \sum_{i=0}^{n-1} \lambda_i (\beta - 1)\beta^i - \sum_{i=0}^{n-1} \lambda_i x_i \beta^i = Q - X \qquad (2.7)$$

Hence,

$$-X = \overline{X} - Q = \overline{X} + (-Q). \qquad (2.8)$$

In other words, the additive inverse of X can be formed by adding the additive inverse of Q to the complement $\overline{X}$.

Example 2.4

In the two's complement system, $Q = -ulp$ and therefore $-X = \overline{X} + ulp$. In the nega-binary system, $Q = (\cdots, 1, 1, 1, 1) = -(\cdots, 0, 1, 0, 1)$ and hence, $-Q = (\cdots, 0, 1, 0, 1)$. The additive inverse of $(01011)_{-2} = (-9)_{10}$, for example, is

	16	−8	+4	−2	+1	
$\overline{X}$	1	0	1	0	0	
$-Q \;\; +$	1	0	1	0	1	
	1	1	0	0	1	$= 9_{10}$

where the addition is performed according to the rules of adding nega-binary numbers. This result can be verified by adding the original number to its additive inverse. Again following the rules of nega-binary addition, $(01011)_{-2} + (11001)_{-2} = (00000)_{-2}$.   □

The additive inverse may be employed in subtraction by using the equation

$$X - Y = X + \overline{Y} + (-Q). \qquad (2.9)$$

Applying this equation may require two add operations with carry propagation, which is time-consuming. An alternate expression is

$$X - Y = \overline{\overline{X} + Y}. \qquad (2.10)$$

Two digit-complement operations, and only one addition with carry-propagation, are required here.

2.3 SIGNED-DIGIT NUMBER SYSTEMS

In all fixed-radix systems that we have examined so far, the digit set has been restricted to $\{0, \cdots, r - 1\}$. However, we can allow the following digit set:

$$x_i \in \{\overline{(r-1)}, \overline{(r-2)}, \cdots, \overline{1}, 0, 1, \cdots, (r-1)\}, \qquad (2.11)$$

where $\bar{i}$ equals $-i$ and not $(r-1)-i$ as before. We use here the same notation as before, since this is done commonly in the technical literature. Each digit is either positive or negative, so there is no need for a separate sign digit. The resulting number system is called the *signed-digit* (*SD*) system.

Example 2.5

> For $r = 10$ the allowed digits are $\{\bar{9}, \bar{8}, \cdots, \bar{1}, 0, 1, \cdots, 8, 9\}$ and, if $n = 2$, the range is $\bar{99} \le X \le 99$, which includes 199 numbers. However, with two digits (x_1, x_0) each having 19 possibilities, there are $19^2 = 361$ representations, and hence some numbers have more than one representation. The number system is therefore redundant. For example, $(01) = (1\bar{9}) = 1$; $(0\bar{2}) = (\bar{1}8) = -2$. The representation of 0, however, is unique, and so is the representation of 10. Out of the 361 representations, $361 - 199 = 162$ are redundant, and thus there is 81% redundancy. The reader can verify that each number in this range has at most two representations.   □

As will become apparent later on, adding some redundancy in a number system can be very beneficial. On the other hand, a high level of redundancy might be too costly, since a larger digit set requires a larger number of bits to represent each digit. We may reduce the amount of redundancy by restricting the digit set to

$$x_i \in \{\bar{a}, \overline{a-1}, \ldots, \bar{1}, 0, 1, \ldots, a\} \quad \text{with} \quad \left\lceil \frac{r-1}{2} \right\rceil \le a \le r-1 \quad (2.12)$$

where the ceiling $\lceil x \rceil$ of a number x is the smallest integer that is larger than or equal to x. At least r different digits are needed to represent a number in a radix r system and with $\bar{a} \le x_i \le a$ we have $2a + 1$ digits. Therefore, the inequality $2a + 1 \ge r$ must be satisfied and the lower bound in inequality (2.12) follows.

Example 2.6

> For $r = 10$ the range of a is $5 \le a \le 9$. If we select $a = 6$, then for $n = 2$ there are 133 numbers in the range $\bar{66} \le X \le 66$. Each digit has 13 possible values for a total of $13^2 = 169$ representations; i.e., there is a 27% redundancy. Notice that 1 now has only one representation, namely (01). $(1\bar{9})$ is not a valid representation, since $\bar{9}$ is illegal. However, 4 still has two representations: (04) and $(1\bar{6})$.   □

SD representations are useful when developing algorithms for multiplication and division, as described in Chapters 6 and 7. However, the original motivation for introducing *SD* numbers was to eliminate carry propagation chains in addition and subtraction. Consider the following operation:

$$(x_{n-1}, \ldots, x_0) \pm (y_{n-1}, \ldots, y_0) = (s_{n-1}, \ldots, s_0)$$

We want to break the carry chains by having the sum digit s_i depend only on the four operand digits x_i, y_i, x_{i-1} and y_{i-1}. If this can be achieved, then the addition time becomes independent of the length of the operands. An addition algorithm that can achieve this independence consists of two steps:

Step 1: Compute an interim sum u_i and a carry digit c_i:

$$u_i = x_i + y_i - rc_i$$

where

$$c_i = \begin{cases} 1 & \text{if } (x_i + y_i) \geq a \\ \bar{1} & \text{if } (x_i + y_i) \leq \bar{a} \\ 0 & \text{if } |x_i + y_i| < a \end{cases} \qquad (2.13)$$

Step 2: Calculate the final sum $s_i = u_i + c_{i-1}$.

Example 2.7

If we select $a = 6$ for $r = 10$, then $x_i \in \{\bar{6}, \cdots, 0, 1, \cdots, 6\}$. Step 1 in the above addition algorithm then becomes $u_i = (x_i + y_i) - 10c_i$ and

$$c_i = \begin{cases} 1 & \text{if } (x_i + y_i) \geq 6 \\ \bar{1} & \text{if } (x_i + y_i) \leq \bar{6} \\ 0 & \text{otherwise} \end{cases}$$

Thus, instead of performing the addition of two decimal numbers, such as 4536 and 1466, in the conventional decimal number system

```
        4  5  3  6
   +    1  4  6  6
   ───────────────
        6  0  0  2
```

in which the carry propagates from the least significant digit to the most significant one, we have no carry propagation chain in

```
          4  5  3  6
   +      1  4  6  6
   ─────────────────────
      0   1  1  1        c_i
   +      5  1̄  1̄  2     u_i
   ─────────────────────
      6   0  0  2        s_i
```

Note that the carry bits were shifted to the left to simplify the execution of the second step of the algorithm.   □

In the last example the operands were selected so that they could be viewed either as conventional decimal numbers or as *SD* decimal numbers with the digit set $\{\overline{6}, \cdots, 0, 1, \cdots, 6\}$. In general, a conventional decimal number may use digits like 7, 8, and 9, which are out of the allowed digit set. The previous addition algorithm can be used for converting a conventional decimal number to *SD* form by artificially considering each digit as the sum $(x_i + y_i)$ in Equation (2.13).

Example 2.8

To find the *SD* representation of the decimal number 27956 we apply the previous algorithm, resulting in

$x_i + y_i$		2	7	9	5	6
c_i	0	1	1	0	1	
u_i		2	$\overline{3}$	$\overline{1}$	5	$\overline{4}$
s_i		3	$\overline{2}$	$\overline{1}$	6	$\overline{4}$

To convert a number in *SD* representation to conventional representation one can subtract the digits with negative weight from the digits with positive weight. For $3\overline{2}\overline{1}6\overline{4}$ we obtain

$$
\begin{array}{r}
3 \quad 0 \quad 0 \quad 6 \quad 0 \\
- \quad 0 \quad 2 \quad 1 \quad 0 \quad 4 \\
\hline
2 \quad 7 \quad 9 \quad 5 \quad 6
\end{array}
$$

$\square$

To guarantee that no new carry will be generated, the sum digit s_i, calculated from $u_i + c_{i-1}$, must satisfy $|s_i| \leq a$. Since $|c_{i-1}| \leq 1$, the condition $|u_i| \leq a - 1$ has to be satisfied for all possible values of x_i and y_i. For example, the largest value that $x_i + y_i$ can assume is $2a$, for which $c_i = 1$ and $u_i = 2a - r$. The inequality $u_i = 2a - r \leq a - 1$ is clearly satisfied, since $a \leq r - 1$. However, if $x_i + y_i = a$, which is the smallest value for which c_i is still 1, then $u_i = a - r < 0$. Substituting $|u_i| = r - a$ into $|u_i| \leq (a - 1)$ yields the inequality $2a \geq r + 1$. Hence, the selected digit set must satisfy

$$\left\lceil \frac{r+1}{2} \right\rceil \leq a \leq r - 1. \tag{2.14}$$

We have considered so far only the two extreme values of $x_i + y_i$ for which $c_i = 1$. However, the reader can verify that for all other possible values of $x_i + y_i$, the condition $|u_i| \leq a - 1$ is satisfied if $a \geq \left\lceil \frac{r+1}{2} \right\rceil$.

Example 2.9

SD decimal numbers must satisfy $a \geq 6$ to guarantee that no new carries will be generated in the previous algorithm. $\square$

$x_i y_i$	00	01	0$\bar{1}$	11	$\bar{1}\bar{1}$	1$\bar{1}$
c_i	0	1	$\bar{1}$	1	$\bar{1}$	0
u_i	0	$\bar{1}$	1	0	0	0

TABLE 2.1   The rules for adding binary *SD* numbers.

2.4 BINARY *SD* NUMBERS

For $r = 2$, there is only one possible digit set; namely, $\{\bar{1}, 0, 1\}$. In other words, a must equal 1. The interim sum and carry in the addition algorithm from Section 2.3 are

$$u_i = (x_i + y_i) - 2c_i$$

and

$$c_i = \begin{cases} 1 & \text{if } (x_i + y_i) \geq 1 \\ \bar{1} & \text{if } (x_i + y_i) \leq \bar{1} \\ 0 & \text{if } (x_i + y_i) = 0. \end{cases}$$

These rules are summarized in Table 2.1. This table does not include the combinations $x_i y_i = 10$, $x_i y_i = \bar{1}0$, and $x_i y_i = \bar{1}1$, since the addition $x_i + y_i$ is a commutative operation. Note that in the binary case the condition $a \geq \lceil \frac{r+1}{2} \rceil = 2$ cannot be satisfied, and consequently there is no guarantee that a new carry will not be generated in the second step of the algorithm.

Still, if the operands to be added do not contain the digit $\bar{1}$, new carries will not be generated. Consider, for example, the addition of the following two numbers, which, in the conventional representation, will generate a carry that will propagate from the least significant position to the most significant position:

$$
\begin{array}{ccccccc}
 & 1 & 1 & \cdots & 1 & 1 & \\
+ & 0 & 0 & \cdots & 0 & 1 & \\
\hline
1 & 1 & 1 & \cdots & 1 & & c_i \\
 & \bar{1} & \bar{1} & \cdots & \bar{1} & 0 & u_i \\
\hline
 & 1 & 0 & 0 & \cdots & 0 & 0 & s_i \\
\end{array}
$$

Here, no carry propagation chain exists. However, if *SD* numbers with $\bar{1}$ digits are added, new carries may occur. For example, if $x_{i-1} y_{i-1} = 01$, then $c_{i-1} = 1$; and if $x_i y_i = 0\bar{1}$, then $u_i = 1$, yielding $s_i = u_i + c_{i-1} = 1 + 1$. Thus, a new carry is generated, as illustrated in the following addition:

$$
\begin{array}{rccccccc}
 & 0 & \bar{1} & 1 & \bar{1} & 1 & 1 & \\
+ & 1 & 0 & 0 & \bar{1} & 0 & 1 & \\
\hline
1 & \bar{1} & 1 & \bar{1} & 1 & 1 & & c_i \\
 & \bar{1} & 1 & \bar{1} & 0 & \bar{1} & 0 & u_i \\
\hline
 & * & * & * & 1 & 0 & 0 & s_i \\
\end{array}
$$

The stars indicate positions where new carries are generated and must be allowed to propagate.

Examining the rules in Table 2.1, one can verify that the combination $c_{i-1} = u_i = 1$ occurs when $x_i y_i = 0\bar{1}$ and $x_{i-1} y_{i-1}$ equals either 11 or 01. We can avoid setting $u_i = 1$ in these cases by selecting $c_i = 0$ and therefore making u_i equal $\bar{1}$. We should not, however, change the entry for $x_i y_i = 0\bar{1}$ in Table 2.1 to read $c_i = 0$ and $u_i = \bar{1}$, since for $x_{i-1} y_{i-1} = \bar{1}\bar{1}$, which results in $c_{i-1} = \bar{1}$, we still have to set $c_i = \bar{1}$ and $u_i = 1$. Similarly, the combination $c_{i-1} = u_i = \bar{1}$ occurs when $x_i y_i = 01$ and $x_{i-1} y_{i-1}$ equals either $\bar{1}\bar{1}$ or $0\bar{1}$. We can avoid setting $u_i = \bar{1}$ by selecting in these cases (and in these cases only) $c_i = 0$ and therefore $u_i = 1$. In summary, we can ensure that no new carries will be generated by examining the two bits to the right $x_{i-1} y_{i-1}$ when determining u_i and c_i, arriving at the rules shown in Table 2.2. Observe that we can still calculate c_i and u_i for all bit positions in parallel.

$x_i y_i$	00	01	01	$0\bar{1}$	$0\bar{1}$	11	$\bar{1}\bar{1}$
$x_{i-1}\, y_{i-1}$	–	neither is $\bar{1}$	at least one is $\bar{1}$	neither is $\bar{1}$	at least one is $\bar{1}$	–	–
c_i	0	1	0	0	$\bar{1}$	1	$\bar{1}$
u_i	0	$\bar{1}$	1	$\bar{1}$	1	0	0

TABLE 2.2   Modified rules for adding binary *SD* numbers (8).

Example 2.10

Repeating the previous example we obtain

$$
\begin{array}{rccccccc}
 & 0 & \bar{1} & 1 & \bar{1} & 1 & 1 & \\
+ & 1 & 0 & 0 & \bar{1} & 0 & 1 & \\
\hline
0 & 0 & 0 & \bar{1} & 1 & 1 & & c_i \\
 & 1 & \bar{1} & 1 & 0 & \bar{1} & 0 & u_i \\
\hline
 & 1 & \bar{1} & 0 & 1 & 0 & 0 & s_i \\
\end{array}
$$

Note that direct summation of the two operands will result in $1\bar{1}1\bar{1}00$. These, and also 010100, are equivalent, and all represent the value 20_{10}.

□

Binary *SD* numbers are particularly useful in the development of fast algorithms for multiplication and division, which are discussed in Chapters 6 and 7. In these cases we will be interested in *minimal SD representations;* i.e., representations that include a minimal number of nonzero digits. Nonzero digits will correspond to add/subtract operations, and the number of these should be minimized while zero digits will correspond to shift-only operations.

Example 2.11

For $X = 7$ we have the following representations:

$$
\begin{array}{ccccc}
 & 8 & 4 & 2 & 1 \\
\hline
 & 0 & 1 & 1 & 1 \\
 & 1 & \bar{1} & 1 & 1 \\
 & 1 & 0 & \bar{1} & 1 \\
 & 1 & 0 & 0 & \bar{1} \\
1 & \bar{1} & \bar{1} & 1 & 1 \\
 & & \vdots & &
\end{array}
$$

Out of these, $100\bar{1}$ is the minimal representation. The canonical recoding algorithm generates minimal *SD* representations of given binary numbers and is presented in Chapter 6. ☐

Eliminating carry propagation when adding binary numbers can speed up operations like multiplication and division, whose execution usually includes a large number of add/subtract operations. Two concerns arise when *SD* representations of binary numbers are used in an arithmetic unit. The first one is the exact encoding of three values, namely 0, 1 and -1, using binary signals. The second one is the need to convert the result in *SD* representation to its conventional two's complement representation. Out of the $4! = 24$ possible ways to encode the three values of a binary signed digit x using two bits, x^h and x^l (h and l for high and low, respectively), only nine are distinct encodings under permutation and logical negation. Out of these nine encodings two have been used in practice. These are shown in Table 2.3.

x	Encoding 1 $x^h \; x^l$	Encoding 2 $x^h \; x^l$
0	0 0	0 0
1	0 1	0 1
$\bar{1}$	1 0	1 1

TABLE 2.3 Two encodings for binary *SD* numbers.

Encoding 2 can be viewed as a two's complement representation of the signed digit x. Encoding 1 is sometimes preferable since it satisfies the simple relation

$$x = x^l - x^h$$

and consequently, the combination 11 has a valid numerical value of 0. This encoding also simplifies the conversion of a number from the SD to the two's complement representation. This conversion is done by subtracting the sequence $x^h_{n-1}, x^h_{n-2}, \cdots, x^h_0$ from the sequence $x^l_{n-1}, x^l_{n-2}, \cdots, x^l_0$, using two's complement arithmetic.

There exists another conversion algorithm whose implementation requires a circuit simpler than a complete binary adder ([7], [9]). In this algorithm the binary signed digits are examined from right to left, one digit at a time. The algorithm removes all occurrences of $\bar{1}$ digits and "forwards" the negative sign to the most significant bit, the only bit with a negative weight in the two's complement representation. The rightmost $\bar{1}$ digit is replaced by a 1 and the negative sign is forwarded to the left, replacing 0's by 1's until a 1 is reached, which "consumes" the negative sign and is replaced by a 0. If a 1 is not reached then the 0 in the most significant position is turned into a 1, becoming the negative sign bit of the two's complement representation. If a second $\bar{1}$ is encountered before a 1 is, it is replaced by a 0 and the forwarding of the negative sign continues. The negative sign is forwarded with the aid of a a "borrow" bit which equals 1 as long as a $\bar{1}$ is being forwarded, and equals 0 otherwise. The rules of this algorithm are summarized in Table 2.4, where y_i is the ith digit of the SD number, z_i is the ith bit of the corresponding two's complement representation, c_i is the previous "borrow" and c_{i+1} is the next "borrow." For the least significant digit we assume $c_0 = 0$. This algorithm performs the inverse operation to that performed by the *canonical recoding* algorithm presented in Section 6.1. It satisfies the arithmetic equation

$$y_i - c_i = z_i - 2c_{i+1}.$$

y_i	c_i	z_i	c_{i+1}
0	0	0	0
0	1	1	1
1	0	1	0
1	1	0	0
$\bar{1}$	0	1	1
$\bar{1}$	1	0	1

TABLE 2.4   An algorithm for converting SD to two's complement representation.

By rearranging terms in the above equation we can obtain the arithmetic equation satisfied by the canonical recoding algorithm:

$$z_i + c_i = y_i + 2c_{i+1}$$

Example 2.12

To convert the *SD* representation of the number -10_{10} to two's complement we apply the previous algorithm, resulting in

y_i		0	$\bar{1}$	0	$\bar{1}$	0
c_i	1	1	1	1	0	0
z_i		1	0	1	1	0

Since the range of representable numbers in the *SD* method is almost double that of the two's complement method, an n-digit *SD* number must be converted to an $(n+1)$-bit two's complement representation as illustrated below.

y_i		0	1	0	1	0	$\bar{1}$
c_i	0	0	0	0	1	1	0
z_i		0	1	0	0	1	1

Without the extra bit position the number 19 would be converted to -13.

$\square$

The use of Encoding 2 in Table 2.3 also has some advantages. Since the value of the operand digit x_i is given by $-2x_i^h + x_i^l$ there are bits, namely x_i^l and x_{i-1}^h, in two adjacent digit positions which have the same weight. We can then regroup the bits x_i^l and x_{i-1}^h to form a new digit $\hat{x}_i$. Performing such equal-weight grouping [6] allows us to derive new addition rules for $\hat{x}_i$ and $\hat{y}_i$ which may lead to simpler implementations.

Note that in Table 2.5 the bits x_{i-1}^l and y_{i-1}^l are needed only when $\hat{x}_i \hat{y}_i = 1$ but not when $\hat{x}_i \hat{y}_i = \bar{1}$. This may lead to simpler implementation compared to Table 2.5 where the information about the previous digits is required also for the case $x_i y_i = \bar{1}$.

$\hat{x}_i \hat{y}_i$	00	01	01	01	0$\bar{1}$	11	$\bar{1}\bar{1}$
$x_{i-1}^h y_{i-1}^h$	–	both are 0	both are 0	at least one is 1	–	–	–
$x_{i-1}^l y_{i-1}^l$	–	at least one is 1	both are 0	–	–	–	–
c_i	0	1	0	1	0	1	$\bar{1}$
u_i	0	$\bar{1}$	1	$\bar{1}$	$\bar{1}$	0	0

TABLE 2.5   Rules for adding binary *SD* numbers with equal-weight grouping (6).

2.5 EXERCISES

2.1. Find the fixed-point representations of the two values $(41.25)_{10}$ and $(-41.25)_{10}$ in the following radix-r number systems. Each number consists of k integer digits and m fraction digits.
(a) Signed-digit representation with $r = 4$, $k = 4$, $m = 2$, and the digit set $(\bar{2}, \bar{1}, 0, 1, 2)$. At least one negative digit must appear in the representation.
(b) A Nega-decimal system with $r = -10$, $k = 3$, and $m = 2$.

2.2. Given the value $(-14)_{10}$, the word length $n = 6$, and the digit set $(\bar{1}, 0, 1)$, find all the possible six-digit SD representations of the given value. Indicate the minimal representation. Is this representation unique?

2.3. Verify that the representation of 0 in an SD number system is unique.

2.4. Prove that any fixed-radix number system $< n, \beta, \Lambda >$ is nonredundant.

2.5. Calculate the difference between the two numbers 0010 and 0101 (in the conventional binary system) in two ways: first, in the traditional way by adding the two's complement of 0101 to 0010, then by using Equation (2.10). Is there an advantage to using one of the two methods?

2.6. Show that if $a \geq \left\lceil \frac{r+1}{2} \right\rceil$ no new carry will be generated when adding SD numbers.

2.7. Find all the values of the radix r for which the two-step algorithm for adding SD numbers will not generate new carries if and only if the maximum redundancy is followed.

2.8. Show that in the negative-radix system with $\beta > 2$ the additive inverse of Q is $-Q = (\cdots, 2, 2, 2, 1)$. Find the additive inverse of 08019 in the nega-decimal system. Verify your result by adding it to 08019.

2.9. Can modified rules such as those in Table 2.2 be derived for $r > 2$ so that less redundancy (e.g., $a = \left\lceil \frac{r-1}{2} \right\rceil$) will be needed?

2.10. Are the modified rules for adding binary SD numbers in Table 2.2 unique? In other words, can you suggest another set of rules that will guarantee that no new carries will be generated in the second step of the addition?

2.11. An imaginary-radix number system can be defined as follows. Let the radix r have the form $r = j\sqrt{\beta}$ where $j = \sqrt{-1}$ and β is a positive integer. Let the digit set be $\{0, 1, ..., \beta - 1\}$. Show that all even-position digits represent a real number Y in the negative-radix $(-\beta)$ system and all odd-position digits represent an imaginary number $j\sqrt{\beta}Z$ in the same negative-radix number system. As a result, we can write $X = Y + j\sqrt{\beta}Z$, representing a complex number by a single sequence. Therefore, instead of performing four multiplications and two summations:

$$(Y_1 + j\sqrt{\beta}Z_1) \times (Y_2 + j\sqrt{\beta}Z_2) = (Y_1 Y_2 - \beta Z_1 Z_2) + j\sqrt{\beta}(Y_1 Z_2 + Y_2 Z_1)$$

we can directly multiply two complex numbers X_1 and X_2. Will the use of this imaginary-radix number system speed up the multiplication of complex numbers?

2.12. Write the Boolean equations for a circuit implementing the conversion algorithm in Table 2.4 using encoding 1 from Table 2.3. Point out the similarities between the resulting circuit and a full-adder (FA, see Section 5.1). Discuss the possibility of employing any speedup techniques used for binary addition.

2.13. Show that the algorithm, summarized in Table 2.4, for converting an SD to a two's complement representation of a binary number, can be performed by forming the sequence $0, |y_{n-1}|, |y_{n-2}|, \cdots, |y_1|, |y_0|$ and than performing a bit-wise exclusive-OR operation with the sequence $c_n, c_{n-1}, c_{n-2}, \cdots, c_1, 0$ which is obtained according to the rules in Table 2.4. This version of the conversion algorithm was presented in [7].

2.14. A hybrid SD number system was presented in [5]. In this system only some digit positions are signed while the rest remain unsigned. Develop two sets of rules for adding the corresponding digits of two hybrid numbers, similar to those in Table 2.2. The first table will indicate the rules for selecting the carry c_i and intermediate sum u_i for two signed digits x_i and y_i with a_{i-1} and b_{i-1} being the lower order two unsigned bits. The second table will indicate the rules for selecting the carry c_i and intermediate sum e_i for two unsigned digits a_i and b_i with an incoming carry c_{i-1} that can assume the values 0, 1, and -1. Show that if d is the longest distance between neighboring signed digits then the maximum carry propagation chain is of length $(d+1)$.

2.15. Show that the addition rules in Table 2.5 guarantee that no new carries will be generated in the second step of the addition.

2.6 REFERENCES

[1] A. AVIZIENIS, "Signed-digit number representations for fast parallel arithmetic," *IRE Trans. on Elec. Computers, EC-10* (Sept. 1961), 389-400.

[2] H. L. GARNER, "Number systems and arithmetic," in *Advances in Computers*, vol. 6, F. L. Alt and M. Rubinoff (Eds.), Academic, New York, 1965, pp. 131-194.

[3] I. KOREN and Y. MALINIAK, "On classes of positive, negative and imaginary radix number systems," *IEEE Trans. on Computers, C-30* (May 1981), 312-317.

[4] B. PARHAMI, "Generalized signed-digit number systems: A unifying framework for redundant number representations," *IEEE Trans. on Computers, 39* (Jan. 1990), 89-98.

[5] D.S. PHATAK and I. KOREN, "Hybrid Signed-Digit Number Systems: A Unified Framework for Redundant Number Representations with Bounded Carry Propagation Chains," *IEEE Trans. on Computers, 43* (August 1994), 880-891.

[6] D.S. PHATAK, T. GOFF, and I. KOREN, "Constant-time addition and simultaneous format conversion based on redundant binary representations," *IEEE Trans. on Computers, 50*, (2001).

[7] H. R. SRINIVAS and K. K. PARHI, "A fast VLSI adder architecture," *IEEE J. of Solid-State Circuits, 27* (May 1992), 761-767.

[8] N. TAKAGI, H. YASUURA, and S. YAJIMA, "High speed VLSI multiplication algorithm with a redundant binary addition tree," *IEEE Trans. on Computers, 34* (Sept. 1985), 789-796.

[9] S. M. YEN, C. S. LAIH, C. H. CHEN, and J. Y. LEE, "An efficient redundant-binary number to binary number converter," *IEEE J. of Solid-State Circuits, 27* (Jan. 1992), 109-112.

3

SEQUENTIAL ALGORITHMS FOR MULTIPLICATION AND DIVISION

This chapter presents the basic sequential algorithms for multiplication, division, and square root extraction. Algorithms for high-speed multiplication are described in Chapter 6. Chapters 7, and 8 include algorithms for fast division and high-speed calculation of square roots.

3.1 SEQUENTIAL MULTIPLICATION

Let the multiplier and multiplicand be denoted by X and A, respectively, with the following sequences of digits:

$$X = x_{n-1}x_{n-2} \cdots x_1 x_0 \;, \qquad A = a_{n-1}a_{n-2} \cdots a_1 a_0$$

where x_{n-1} and a_{n-1} are the sign digits in either the signed-magnitude or the complement methods.

The sequential algorithm for multiplication consists of $n-1$ steps where in step j the multiplier bit x_j is examined and the product $x_j A$ is added to the previously accumulated partial product, denoted by $P^{(j)}$. The appropriate expression for this recursive procedure is

$$P^{(j+1)} = (\ P^{(j)} + x_j \cdot A) \cdot 2^{-1} \;; \qquad j = 0, 1, 2, \cdots, n-2 \qquad (3.1)$$

where in the first step $P^{(0)} = 0$. Multiplying the sum $(P^{(j)} + x_j A)$ by 2^{-1} shifts it by one position to the right, to align $P^{(j+1)}$ before adding the next product $x_{j+1}A$. This alignment is necessary, since the weight of x_{j+1} is double that of

x_j. To prove that the above procedure calculates the product of A and X, we repeatedly substitute into the recursive Equation (3.1), yielding

$$
\begin{aligned}
P^{(n-1)} &= (P^{(n-2)} + x_{n-2} \cdot A) \cdot 2^{-1} \\
&= \left((P^{(n-3)} + x_{n-3} \cdot A) \cdot 2^{-1} + x_{n-2} \cdot A \right) \cdot 2^{-1} = \cdots \\
&= \left(x_{n-2} 2^{-1} + x_{n-3} 2^{-2} + \cdots + x_0 2^{-(n-1)} \right) \cdot A \\
&= (\sum_{j=0}^{n-2} x_j \, 2^{-(n-1-j)} \,) \cdot A = 2^{(n-1)} \left(\sum_{j=0}^{n-2} x_j \, 2^j \right) \cdot A
\end{aligned}
$$

If both operands are positive (i.e., $x_{n-1} = a_{n-1} = 0$), the product U is obtained from

$$
U = 2^{n-1} \cdot P^{(n-1)} = (\sum_{j=0}^{n-2} x_j \, 2^j) \cdot A = X \cdot A \tag{3.2}
$$

The result is a product consisting of $2(n-1)$ bits for its magnitude. To prove this, note that the maximum value of U is obtained when A and X assume their maximum value. Therefore,

$$
U_{max} = (2^{n-1} - 1)(2^{n-1} - 1) = 2^{2n-2} - 2^n + 1 = 2^{2n-3} + (2^{2n-3} - 2^n + 1) \tag{3.3}
$$

Since the last term in Equation (3.3) is positive for $n \geq 3$, the following inequality holds:

$$
2^{2n-3} < U_{max} < 2^{2n-2} \, ; \qquad n \geq 3 \tag{3.4}
$$

Thus, $(2n - 2)$ bits are required to represent the value, producing a total of $(2n - 1)$ bits when added to the sign bit.

For signed-magnitude numbers we multiply the two magnitudes using the above algorithm and generate the sign of the result separately (it is positive if both operands have the same sign and negative otherwise). For two's and one's complement representations we should distinguish between multiplication with a negative multiplicand A and multiplication with a negative multiplier X. If only the multiplicand is negative, there is no need to change the previous algorithm. We only to add some multiple of a negative number that is represented in either two's or one's complement. This is illustrated in the next example.

Example 3.1

In the following multiply operation, the multiplicand A is a negative number represented in the two's complement method, while the multiplier X is positive. Both are four bits long and the final product therefore has seven bits, including the sign bit. In an arithmetic unit for 4-bit operands, all registers are four bits long, and consequently a double-length register is required for storing the final product. The vertical line in the table below separates the most significant half of the product, which can be stored in a single-length register (four bits long), from the least significant half, which can be stored in a second single-length register.

A		1	0	1	1					-5
X	$\times$	0	0	1	1					3
$P^{(0)} = 0$		0	0	0	0					
$x_0 = 1 \Rightarrow$ Add A	$+$	1	0	1	1					
		1	0	1	1					
Shift		1	1	0	1	1				
$x_1 = 1 \Rightarrow$ Add A	$+$	1	0	1	1					
		1	0	0	0	1				
Shift		1	1	0	0	0	1			
$x_2 = 0 \Rightarrow$ Shift only		1	1	1	0	0	0	1		-15

The three bits of the multiplier, x_2, x_1, and x_0, are examined one bit at a time, starting with the least significant bit x_0. An add-and-shift or shift-only operation is then performed accordingly. The final result is negative and is properly represented in two's complement. Note that the partial product bits in the least significant half do not participate in the add operation, and that all four bit positions in the first register (holding the most significant half of the final product) are utilized. However, only three bit positions in the second register are utilized, leaving the least significant bit position unused. This need not necessarily be the final arrangement. The three bits in the second register can afterwards be stored in the three rightmost positions, and the sign bit of the second register can then be set according to one of the following two possibilities: (1) Always set the sign bit to 0, irrespective of the sign of the product, since it is the least significant part of the result; (2) Set the sign bit equal to the sign bit of the first register. Another possible arrangement is to use all four bit positions in the second register for the four least significant bits of the product, use the rightmost two bit positions in the first register, and insert two copies of the sign bit into the remaining bit positions. □

The situation, however, is different when the multiplier is negative. Here, we consider each bit separately, and the sign bit (which has a negative weight) cannot be treated in the same way as the other bits. First consider two's complement numbers, which satisfy

$$X = -x_{n-1}\, 2^{n-1} + \widetilde{X} \tag{3.5}$$

where $\widetilde{X} = \sum_{j=0}^{n-2} x_j 2^j$.

If the sign bit of the multiplier in the previously presented procedure is ignored, then the final result U satisfies

$$U = \widetilde{X} \cdot A = (X + x_{n-1} \cdot 2^{n-1}) \cdot A = X \cdot A + A \cdot x_{n-1} \cdot 2^{n-1}. \tag{3.6}$$

The term $X \cdot A$ is the desired product and hence, if $x_{n-1} = 1$, the following correction is necessary:

$$X \cdot A = U - A \cdot x_{n-1} \cdot 2^{n-1} \tag{3.7}$$

In other words, if $x_{n-1} = 1$, we must subtract the multiplicand A from the most significant half of U.

Example 3.2

The multiplier and multiplicand in this example are both negative numbers in the two's complement representation:

A			1	0	1	1				-5
X	$\times$		1	1	0	1				-3
$x_0 = 1 \Rightarrow$ Add A			1	0	1	1				
Shift			1	1	0	1	1			
$x_1 = 0 \Rightarrow$ Shift only			1	1	1	0	1	1		
$x_2 = 1 \Rightarrow$ Add A	$+$		1	0	1	1				
			1	0	0	1	1	1		
Shift			1	1	0	0	1	1	1	
$x_3 = 1 \Rightarrow$ Correct	$+$		0	1	0	1				
			0	0	0	1	1	1	1	$+15$

In the correction step, the subtraction of the multiplicand is performed by adding its two's complement. □

Similarly, when multiplying one's complement numbers, which satisfy

$$X = -x_{n-1}(2^{n-1} - ulp) + \widetilde{X} \tag{3.8}$$

then,

$$X \cdot A = U - x_{n-1} \cdot 2^{n-1} \cdot A + x_{n-1} \cdot ulp \cdot A. \tag{3.9}$$

Thus, if $x_{n-1} = 1$, we start with $P^{(0)} = A$, which takes care of the second correction term, namely, $x_{n-1} \cdot ulp \cdot A$, and at the end of the process we subtract the first correction term, $A \cdot x_{n-1} \cdot 2^{n-1}$.

Example 3.3

The product of 5 and -3 in one's complement representation is

A			0	1	0	1				5
X	$\times$		1	1	0	0				-3
$x_3 = 1 \Rightarrow P^{(0)} = A$			0	1	0	1				
$x_0 = 0 \Rightarrow$ Shift			0	0	1	0	1			
$x_1 = 0 \Rightarrow$ Shift			0	0	0	1	0	1		
$x_2 = 1 \Rightarrow$ Add A	$+$		0	1	0	1				
			0	1	1	0	0	1		
Shift			0	0	1	1	0	0	1	
$x_3 = 1 \Rightarrow$ Correct	$+$		1	0	1	0	1	1	1	
			1	1	1	0	0	0	0	-15

As in the previous example, the subtraction of the (first) correction term is accomplished by adding its one's complement. However, unlike the previous example, the one's complement has to be expanded to double size using the sign digit (see Section 1.6). This implies that a double-length binary adder is needed. $\square$

3.2 SEQUENTIAL DIVISION

Division is the most complex of the four basic arithmetic operations and, consequently, the most time-consuming. Unlike the other three operations, division, in general, has a result consisting of two components. Given a dividend X and a divisor D, a quotient Q and a remainder R have to be calculated so as to satisfy

$$X = Q \cdot D + R \quad \text{with} \quad R < D. \tag{3.10}$$

We will assume at first, for simplicity, that the operands X and D and the results Q and R are positive numbers.

In many fixed-point arithmetic units, a double-length product is available after a multiply operation, and we wish to allow the use of this result in a subsequent divide operation. Thus, X may occupy a double-length register, while all other operands are stored in single-length registers. Consequently, we have to make sure that the resulting quotient Q is smaller than or equal to the largest number that we can store in a single-length register. If n is the number of bits in a single-length register, then every single-length integer is smaller than 2^{n-1}. Therefore, to ensure that the quotient is a single-length integer (i.e., the inequality $Q < 2^{n-1}$ is satisfied), we must require that

$$X < 2^{n-1} D.$$

If this condition is not satisfied, an *overflow* indication should be produced by the arithmetic unit. One should be aware that the above condition can always be satisfied by preshifting one of the operands X or D (or both). This preshifting is especially simple to apply when the operands are floating-point numbers. Another condition that has to be checked is that $D \neq 0$. If this is not the case, a *divide by zero* indication should be generated by the arithmetic unit. Unlike the previous condition, no corrective action can be taken when $D = 0$.

The presentation of algorithms for division is simpler when the dividend and divisor, as well as the quotient and remainder, are interpreted as fractions. In this case, the divide overflow condition becomes $X < D$ to ensure that the quotient is a fraction. The division procedure that is presented next assumes that all operands and results are fractions, but is clearly also valid for integers, as will become apparent later on.

To obtain the fractional (positive) quotient $Q = 0.q_1 \cdots q_m$ (where $m = n - 1$), we perform the division as a sequence of subtractions and shifts. In step i of the process the remainder is compared to the divisor D. If the remainder is the larger of the two, then the quotient bit q_i is set to 1. If not, it is set to 0. The equation for the ith step is

$$r_i = 2r_{i-1} - q_i \cdot D ; \qquad i = 1, 2, \cdots, m \tag{3.11}$$

where r_i is the new remainder and r_{i-1} is the previous remainder. The first remainder is $r_0 = X$. Thus, q_i is determined by comparing $2r_{i-1}$ to D. This comparison is the most complicated operation in the division process.

We will now prove that the above procedure indeed calculates the quotient and the final remainder. The remainder in the last step is r_m and repeated substitution of Equation (3.11) yields

$$
\begin{aligned}
r_m &= 2r_{m-1} - q_m \cdot D \\
&= 2(2r_{m-2} - q_{m-1} \cdot D) - q_m \cdot D = \cdots \\
&= 2^m r_0 - (q_m + 2q_{m-1} + \cdots + 2^{m-1}q_1) \cdot D.
\end{aligned}
$$

Substituting $r_0 = X$ and dividing both sides by 2^m results in

$$r_m 2^{-m} = X - (q_1 2^{-1} + q_2 2^{-2} + \cdots + q_m 2^{-m}) \cdot D;$$

hence

$$r_m \, 2^{-m} = X - Q \cdot D \tag{3.12}$$

as required. Note that the true final remainder is $R = r_m 2^{-m}$.

Example 3.4

Let $X = (0.100000)_2 = 1/2$ and $D = (0.110)_2 = 3/4$. The dividend occupies a double-length register. The condition $X < D$ is clearly satisfied.

$r_0 = X$			0	.1	0	0	0	0	0	
$2r_0$		0	1	.0	0	0	0	0		set $q_1 = 1$
Add $-D$	+	1	1	.0	1	0				
$r_1 = 2r_0 - D$		0	0	.0	1	0	0	0		
$2r_1$		0	0	.1	0	0	0			set $q_2 = 0$
$r_2 = \ 2r_1$		0	0	.1	0	0	0			
$2r_2$		0	1	.0	0	0				set $q_3 = 1$
Add $-D$	+	1	1	.0	1	0				
$r_3 = 2r_2 - D$		0	0	.0	1	0				

Note that the generation of $2r_0$ should not result in an overflow indication (multiplying a positive number by 2 should result in a positive number), since the quotient and remainder are within the proper range for the given dividend and divisor. Hence, an extra bit position in the arithmetic unit is needed.

The final results are $Q = (0.101)_2 = 5/8$ and $R = r_m 2^{-m} = r_3 2^{-3} = 1/4 \cdot 2^{-3} = 1/32$. (The precise quotient is the infinite binary fraction $2/3 = 0.1010101 \cdots$.) The quotient and final remainder satisfy the equation $X = Q \cdot D + R = 5/8 \cdot 3/4 + 1/32 = 16/32 = 1/2$. $\qquad\square$

Exactly the same procedure should be followed if the operands and results are integers. In this case we may rewrite Equation (3.10) as follows:

$$2^{2n-2} X_F = 2^{n-1} Q_F \cdot 2^{n-1} D_F + 2^{n-1} R_F \qquad (3.13)$$

where X_F, D_F, Q_F, and R_F are fractions. Dividing Equation (3.13) by 2^{2n-2} yields

$$X_F = Q_F \cdot D_F + 2^{-(n-1)} R_F. \qquad (3.14)$$

The above mentioned condition $X < 2^{n-1} D$, when divided by 2^{2n-2}, now takes the form $X_F < D_F$.

Example 3.5

We repeat the previous example with all operands and results being integers. In this case the double-length dividend is $X = 0100000_2 = 32$, and the divisor is $D = 0110_2 = 6$. The overflow condition $X < 2^{n-1} D$ is tested by comparing the most significant half of X, 0100, to D, 0110. The results of the division are $Q = 0101_2 = 5$ and $R = 0010_2 = 2$. Observe that in the final step of the process the true remainder R is generated and, as can be verified from Equation (3.14), there is no need to further multiply it by $2^{-(n-1)}$. $\qquad\square$

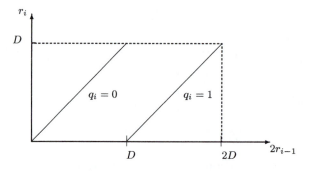

FIGURE 3.1   Restoring division.

The most difficult step in the division procedure is the comparison between the divisor and the remainder to determine the quotient bit. If this is done by subtracting D from $2r_{i-1}$, then in the case of a negative result we set $q_i = 0$, and we must restore the remainder to its previous value. This method is therefore called *restoring division*, and can be diagrammed, as shown in Figure 3.1. Such a diagram is sometimes called a Robertson diagram [7].

This diagram illustrates the fact that if $r_{i-1} < D$, q_i should be selected so as to ensure $r_i < D$. Since $r_0 = X < D$, we are guaranteed to obtain $R < D$. In summary, a division performed by the restoring method uses m subtractions, m shift operations, and an average of $m/2$ restore operations. The latter can be implemented either by adding D or by retaining a copy of the previous remainder, thus avoiding the time penalty involved in the restore operations.

3.3 NONRESTORING DIVISION

An alternative scheme for sequential division is the *nonrestoring* division algorithm, in which the quotient bit is not corrected and the remainder is not restored immediately if it is negative. These corrections are instead postponed to later steps. In the restoring method, if $2r_{i-1} - D$ is negative, the remainder is restored to $2r_{i-1}$. It is then shifted and D is once again subtracted, obtaining $4r_{i-1} - D$. This process is repeated as long as the remainder is negative. In the nonrestoring method we avoid the restore operation, stay with a negative remainder $2r_{i-1} - D < 0$, shift it, and then attempt to correct it by *adding* D, obtaining $2(2r_{i-1} - D) + D = 4r_{i-1} - D$. Thus, this algorithm produces a remainder equal to the one we would generate using restoring division.

Consider now the resulting quotient. To enable the correction of a "wrong" selection of the quotient bit in step i, we must allow the next quotient bit, q_{i+1}, to assume a negative value. In other words, the allowed values for q_i are 1 and $\bar{1}$ where $\bar{1}$ represents -1. If q_i was incorrectly set to 1, resulting in a negative remainder, we would then select $q_{i+1} = \bar{1}$ and *add* D to the remainder. Hence,

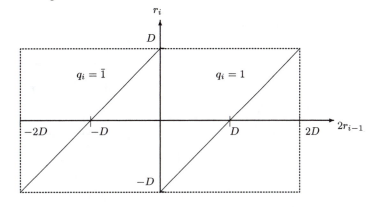

FIGURE 3.2 Nonrestoring division.

instead of $q_i q_{i+1} = 10$ (which is too large), we would get $q_i q_{i+1} = 1\bar{1} = 01$. Further correction, if needed, would be done in the next steps. Consequently, the quotient bit is determined in the nonrestoring scheme by the following rule:

$$q_i = \begin{cases} 1 & \text{if } 2r_{i-1} \geq 0 \\ \bar{1} & \text{if } 2r_{i-1} < 0 \end{cases} \tag{3.15}$$

This rule is simpler (and faster to execute) than the selection rule for restoring division since it requires the comparison of $2r_{i-1}$ to 0 rather than D. The remainder is computed using the same equation

$$r_i = 2r_{i-1} - q_i \cdot D; \tag{3.16}$$

in other words, subtract the divisor D if $2r_{i-1}$ is positive and add it otherwise. The nonrestoring division is diagrammed in Figure 3.2. Here, $|r_{i-1}| < D$ and q_i is selected to ensure $|r_i| < D$. Note that $q_i \neq 0$ and therefore, at each step, either an addition or subtraction is performed. This is not an SD representation, and there is no redundancy in the representation of the quotient in the nonrestoring division. In summary, the nonrestoring method requires exactly m add/subtract and shift operations. Its main advantage is its simpler selection rule.

Example 3.6

Let $X = (0.100)_2 = 1/2$, and $D = (0.110)_2 = 3/4$, as in Example 3.4.

(1)	$r_0 = X$			0	.1	0	0	
(2)	$2r_0$		0	1	.0	0	0	set $q_1 = 1$
(3)	Add $-D$	$+$	1	1	.0	1	0	
(4)	r_1		0	0	.0	1	0	
(5)	$2r_1$		0	0	.1	0	0	set $q_2 = 1$
(6)	Add $-D$	$+$	1	1	.0	1	0	
(7)	r_2		1	1	.1	1	0	
(8)	$2r_2$		1	1	.1	0	0	set $q_3 = \bar{1}$
(9)	Add D	$+$	0	0	.1	1	0	
(10)	r_3		0	0	.0	1	0	

The final remainder is the same as before, and the quotient is $Q = 0.11\bar{1} = 0.101_2 = 5/8$. $\qquad\square$

The nonrestoring division process in the previous example can be represented graphically using a diagram similar to the one depicted in Figure 3.2. The resulting diagram is shown in Figure 3.3. The horizontal lines correspond to the Add $\pm D$ operation in lines 3, 6 and 9 in Example 3.6, and the diagonal lines correspond to the Multiply by 2 operation in lines 2, 5 and 8.

A very important feature of nonrestoring division is that it can easily be extended to two's complement negative numbers. The generalized selection rule for q_i is

$$q_i = \begin{cases} 1 & \text{if } 2r_{i-1} \text{ and } D \text{ have the same sign} \\ \bar{1} & \text{if } 2r_{i-1} \text{ and } D \text{ have opposite signs} \end{cases} \qquad (3.17)$$

Since the remainder changes signs during the process, there is nothing special about a negative dividend X. The following example illustrates the case of a negative divisor in two's complement.

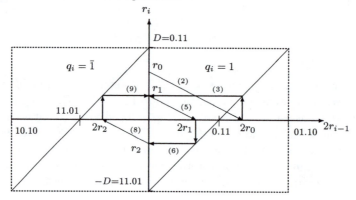

FIGURE 3.3 The nonrestoring division in Example 3.6.

Example 3.7

Let $X = (0.100)_2 = 1/2$ and $D = (1.010)_2 = -3/4$.

$r_0 = X$			0	.1	0	0	
$2r_0$		0	1	.0	0	0	set $q_1 = \bar{1}$
Add D		1	1	.0	1	0	
r_1		0	0	.0	1	0	
$2r_1$		0	0	.1	0	0	set $q_2 = \bar{1}$
Add D	$+$	1	1	.0	1	0	
r_2		1	1	.1	1	0	
$2r_2$		1	1	.1	0	0	set $q_3 = 1$
Add $-D$	$+$	0	0	.1	1	0	
r_3		0	0	.0	1	0	

Finally, $Q = 0.\bar{1}\bar{1}1 = 0.\bar{1}0\bar{1} = -(0.101)_2 = -5/8$, or in two's complement, 1.011. Note that the final remainder is 1/32 and has the same sign as the dividend X. □

By definition, the sign of the final remainder must equal that of the dividend. For example, when dividing 5 by 3 we should obtain a quotient of 1 and a final remainder of 2, and not a quotient of 2 and a final remainder of -1, although this remainder still satisfies $|R| < D$. Consequently, if the sign of the final remainder is different from that of the dividend, a correction of both the final remainder and quotient is needed. This situation, requiring a correction step, arises since the quotient digits in the nonrestoring division algorithm are restricted to $\{1, \bar{1}\}$. The last digit can not be set to 0 and therefore an "even" quotient can not be generated.

Example 3.8

Let $X = (0.101)_2 = 5/8$, and $D = (0.110)_2 = 3/4$. Then

$r_0 = X$			0	.1	0	1	
$2r_0$		0	1	.0	1	0	set $q_1 = 1$
Add $-D$	$+$	1	1	.0	1	0	
r_1		0	0	.1	0	0	
$2r_1$		0	1	.0	0	0	set $q_2 = 1$
Add $-D$	$+$	1	1	.0	1	0	
r_2		0	0	.0	1	0	
$2r_2$		0	0	.1	0	0	set $q_3 = 1$
Add $-D$	$+$	1	1	.0	1	0	
r_3		1	1	.1	1	0	

The final remainder is negative, while the dividend is positive. We must correct the final remainder by adding D to r_3, yielding $1.110+0.110=0.100$, and then correct the quotient:

$$Q_{corrected} = Q - ulp$$

where $Q = 0.111$, and therefore $Q_{corrected} = 0.110_2 = 3/4$.   □

In general, if the final remainder and the dividend have opposite signs, a correction step is needed. If the dividend and divisor have the same sign, then the remainder r_m is corrected by adding D and the quotient is corrected by subtracting ulp. If the dividend and divisor have opposite signs, we subtract D from r_m and correct the quotient by adding ulp.

Another consequence of the fact that 0 is not an allowed digit in non-restoring division, is the need for a correction if a zero remainder is generated in an intermediate step. This case is illustrated in the next example.

Example 3.9

Let $X = (1.101)_2 = -3/8$ and $D = (0.110)_2 = 3/4$. The correct result of this division is $Q = -1/2$ with a zero remainder.

$r_0 = X$				1	.1	0	1	
$2r_0$		1	1	.0	1	0		set $q_1 = \bar{1}$
Add D	$+$	0	0	.1	1	0		
r_1		0	0	.0	0	0		zero remainder
$2r_1$		0	0	.0	0	0		set $q_2 = 1$
Add $-D$	$+$	1	1	.0	1	0		
r_2		1	1	.0	1	0		
$2r_2$		1	0	.1	0	0		set $q_3 = \bar{1}$
Add D	$+$	0	0	.1	1	0		
r_3		1	1	.0	1	0		

Note that although the final remainder r_3 and the dividend X have the same sign, a correction step is needed, since the quotient we get is $Q = 0.\bar{1}1\bar{1} = 0.\bar{1}01_2 = -3/8$ instead of $-1/2$. We must therefore detect the occurrence of a zero intermediate remainder and correct the final remainder (to obtain a zero remainder):

$$r_3(corrected) = r_3 + D = 1.010 + 0.110 = 0.000$$

We have to then correct the quotient $Q = 0.\bar{1}1\bar{1} = 0.\bar{1}01$ by subtracting ulp, yielding $Q_{corrected} = 0.\bar{1}00_2 = -1/2$.   □

3.3.1 Generating a Two's Complement Quotient

The nonrestoring division, as previously presented, generates a quotient that uses the digits 1 and $\bar{1}$ and might therefore be incompatible with the representation used for the dividend and divisor. If X and D are represented in two's complement, then there is a need for a conversion from the above representation to two's complement. We may, in principle, use one of the algorithms presented in Section 2.4 for converting a SD number to its two's complement representation. These algorithms however, require that all the digits of the quotient be known before the conversion can be performed thus increasing the total execution time of the divide operation. We prefer therefore to employ an algorithm that performs the conversion *on the fly*, as the digits of the quotient become available, in a serial fashion from the most to the least significant digit. Such an on-the-fly conversion algorithm from SD to two's complement representation has been presented in [3].

We can however, take advantage of the fact that the quotient digit in the nonrestoring division can assume only the values 1 and $\bar{1}$ (i.e., $q_i \neq 0$) and derive a simpler algorithm that requires a less complex circuit for its implementation. Since the quotient digit can assume only two values, a single bit is sufficient for representing it, and we may assign the digits 0 and 1 to the values $\bar{1}$ and 1, respectively. Let the resulting binary number be denoted by $(0.p_1 \cdots p_m)$ where $p_i = \frac{1}{2}(q_i + 1)$. This number can be converted to two's complement using the following algorithm:

Step 1: Shift the given number one bit position to the left.

Step 2: Complement the most significant bit.

Step 3: Shift a 1 into the least significant position.

The result of this algorithm is the sequence

$$(1 - p_1) \cdot p_2 p_3 \cdots p_m 1.$$

We will now prove that the above sequence, when interpreted as a number in two's complement, has the same numerical value as the original quotient Q. The value of the above sequence in two's complement is

$$-(1 - p_1)2^0 + \sum_{i=2}^{m} p_i 2^{-i+1} + 2^{-m}. \tag{3.18}$$

Substituting $p_i = \frac{1}{2}(q_i + 1)$ yields

$$q_1 2^{-1} - 2^{-1} + \sum_{i=2}^{m} (q_i + 1) 2^{-i} + 2^{-m}$$

$$= q_1 2^{-1} - (2^{-1} - 2^{-m}) + \sum_{i=2}^{m} q_i 2^{-i} + \sum_{i=2}^{m} 2^{-i}.$$

The last term equals $(2^{-1} - 2^{-m})$ and therefore,

$$= q_1 2^{-1} + \sum_{i=2}^{m} q_i 2^{-i} = \sum_{i=1}^{m} q_i 2^{-i} = Q.$$

The above conversion algorithm can be executed in a bit-serial fashion; that is, we can generate the appropriate bit of the quotient, when represented in two's complement, at each step of the nonrestoring division. For example, in the last division with $X = 1.101$ and $D = 0.110$, instead of generating the quotient bits $.1\bar{1}1\bar{1}$, we can generate the bits $(1 - 0).101 = 1.101$. After the correction step we obtain $(Q - ulp) = 1.100$, which is the correct representation of $-1/2$ in two's complement. The same on-the-fly conversion algorithm can be derived from the general SD to two's complement conversion algorithm presented in Section 2.4. This is left as an exercise for the reader.

3.4 SQUARE ROOT EXTRACTION

The conventional "completing the square" method for square root extraction is conceptually similar to the restoring division scheme. Let the given radicand X be a positive fraction, and let $Q = (0.q_1 q_2 \cdots q_m)$ denote its square root. The bits of Q are generated in m steps, one bit per step. We use the notation

$$Q_i = \sum_{k=1}^{i} q_k 2^{-k}$$

for the partially developed root at step i. Thus, $Q_m = Q$. We also denote the remainder in step i by r_i. The next remainder, in general, is calculated from

$$r_i = 2r_{i-1} - q_i \cdot (2Q_{i-1} + q_i 2^{-i}). \tag{3.19}$$

Comparing the above equation to Equation (3.11) suggests that the square root extraction can be viewed as division with a changing divisor, i.e., $\hat{D}_i = (2Q_{i-1} + q_i 2^{-i})$.

In the first step the remainder is the radicand X and $Q_0 = 0$. The performed calculation is therefore

$$r_1 = 2r_0 - q_1(0 + q_1 2^{-1}) = 2X - q_1(0 + q_1 2^{-1}) \tag{3.20}$$

To determine the square root digit q_i in the restoring scheme, a tentative remainder,

$$2r_{i-1} - (2Q_{i-1} + 2^{-i})$$

is calculated. Note that the term $(2Q_{i-1} + 2^{-i})$ is equal to $(q_1.q_2 \cdots q_{i-1} 01$ and is very simple to calculate. If the above tentative remainder is positive, we store its value in r_i and set q_i equal to 1. Otherwise, we set $r_i = 2r_{i-1}$ and $q_i = 0$.

To prove that the above procedure yields the required square root, we repeatedly substitute Equation (3.19) in the expression for r_m, obtaining

$$
\begin{aligned}
r_m &= 2r_{m-1} - q_m(2Q_{m-1} + q_m 2^{-m}) \\
&= 2^2 r_{m-2} - 2q_{m-1}(2Q_{m-2} + q_{m-1}) - q_m(2Q_{m-1} + q_m 2^{-m}) \\
&\quad\vdots \\
&= 2^m \cdot r_0 - 2^m \left[(q_1 2^{-1})^2 + (q_2 2^{-2})^2 + \cdots + (q_m 2^{-m})^2 \right] \\
&\quad - 2^m \left[2q_2 2^{-2} q_1 2^{-1} + \cdots + 2q_m 2^{-m} \sum_{i=1}^{m-1} q_i 2^{-i} \right] \\
&= 2^m X - 2^m \left(\sum_{i=1}^{m} q_i 2^i \right)^2 = 2^m (X - Q^2).
\end{aligned}
$$

Dividing by 2^m results in the expected relation with $r_m 2^{-m}$ as the final remainder.

Example 3.10

Let $X = 0.1011_2 = 11/16 = 176/256$. Its square root is

$r_0 = X$			0	.1	0	1	1
$2r_0$		0	1	.0	1	1	0
$-(0 + 2^{-1})$	$-$	0	0	.1	0	0	0
r_1		0	0	.1	1	1	0	set $q_1 = 1$, $Q_1 = 0.1$
$2r_1$		0	1	.1	1	0	0
$-(2Q_1 + 2^{-2})$	$-$	0	1	.0	1	0	0
r_2		0	0	.1	0	0	0	set $q_2 = 1$, $Q_2 = 0.11$
$2r_2$		0	1	.0	0	0	0	is smaller than $(2Q_2 + 2^{-3})$ $= 1.101$
$r_3 = r_2$		0	1	.0	0	0	0	set $q_3 = 0$, $Q_3 = 0.110$
$2r_3$		1	0	.0	0	0	0	still a positive number
$-(2Q_3 + 2^{-4})$	$-$	0	1	.1	0	0	1
r_4		0	0	.0	1	1	1	set $q_4 = 1$, $Q_4 = 0.1101$

Finally, $Q = 0.1101_2 = 13/16$ and the final remainder is $2^{-4} r_4 = 7/256 = X - Q^2 = (176 - 169)/256$. □

The above procedure is similar to the restoring division algorithm. A method similar to the nonrestoring division algorithm can be employed with the following selection rule for q_i:

$$q_i = \begin{cases} 1 & \text{if } 2r_{i-1} \geq 0 \\ \bar{1} & \text{if } 2r_{i-1} < 0 \end{cases} \tag{3.21}$$

This algorithm is illustrated in the next example.

Example 3.11

Let $X = 0.011001_2 = 25/64$.

$r_0 = X$			0	.0	1	1	0	0	1	
$2r_0$			0	.1	1	0	0	1	0	set $q_1=1$, $Q_1=0.1$
$-(0+2^{-1})$	−		0	.1	0	0	0	0	0	
r_1			0	.0	1	0	0	1	0	
$2r_1$			0	.1	0	0	1	0	0	set $q_2=1$, $Q_2=0.11$
$-(2Q_1+2^{-2})$	−	0	1	.0	1	0	0	0	0	
r_2		1	1	.0	1	0	1	0	0	
$2r_2$		1	0	.1	0	1	0	0	0	set $q_3=\bar{1}$, $Q_3=0.11\bar{1}$
$+(2Q_2-2^{-3})$	+	0	1	.1	0	$\bar{1}$	0	0	0	
r_3		0	0	.0	0	0	0	0	0	

The square root is $Q = 0.11\bar{1} = 0.101_2 = 5/8$.   □

The digits of the square root Q can be converted to two's complement representation by the same method used for the quotient in the nonrestoring division algorithm. Faster algorithms for square root extraction have been developed and implemented. Some of them are introduced in Chapter 7.

3.5 EXERCISES

3.1. Given the following three pairs of binary multiplicand and multiplier:
(i) $+.1001$ and $-.0101$ (ii) $-.1001$ and $+.0101$ (iii) $-.1001$ and $-.0101$.
(a) Represent the numbers in the two's complement form and multiply them. Check your results.
(b) Repeat (a) for the one's complement form.

3.2. Can the sequential multiplication algorithm be modified so that the multiplier bits are examined starting with the most significant bit? What might be a major disadvantage to this modified algorithm?

3.3. Multiply the binary SD numbers $A = 10\bar{1}01$ (the multiplicand) and $X = 0\bar{1}10\bar{1}$ (the multiplier). Perform all intermediate steps in SD arithmetic.

3.4. Given the following three pairs of binary dividend and divisor:
(i) $+.1010$ and $-.1101$ (ii) $-.1010$ and $+.1101$ (iii) $-.1010$ and $-.1101$.

Represent the numbers in the two's complement form and perform the division by the nonrestoring method. The quotient should also be represented in two's complement.

3.5. Devise an algorithm for dividing numbers in one's complement representation. Illustrate your algorithm using the three pairs of numbers in problem 3.4.

3.6. Write the rules for nonrestoring division for decimal fractions. Illustrate the procedure using a positive dividend and positive and negative divisors.

3.7. Explain the need for a correction step in the nonrestoring division if a zero remainder is encountered.

3.8. Show that if the quotient bits q_i $(i = 1, 2, \cdots, m)$ in the nonrestoring division are set according to the rule

$$q_i = \begin{cases} 1 & \text{if the signs of the remainder and divisor agree} \\ 0 & \text{if the signs of the remainder and divisor differ} \end{cases}$$

and subtraction (addition) is performed when $q_1 = 1$ $(q_1 = 0)$, then the correction term $(1 + 2^{-m})$ has to be added to $q_1.q_2 \cdots q_m$ to obtain a quotient represented in two's complement.

3.9. Can the algorithm for converting the quotient bits generated in nonrestoring division into two's complement representation be modified for converting binary *SD* numbers to two's complement? Explain.

3.10. To speed up the nonrestoring division, it has been suggested to allow 0 to be a quotient bit for which no add/subtract operation is needed in order to calculate a new remainder. The modified selection rule is

$$q_i = \begin{cases} 1 & \text{if } 2r_{i-1} \geq D \\ \bar{1} & \text{if } 2r_{i-1} < -D \\ 0 & \text{otherwise} \end{cases}$$

Apply this new algorithm to calculate the quotient of the dividend $X = 0.101$ and the divisor $D = 0.110$. Would you recommend the use of this new algorithm? Explain.

3.11. The on-the-fly conversion algorithm in subsection 3.3.1 is a special case of the *SD*-to-two's complement conversion algorithm presented in Section 2.4. Since just the values 1 and $\bar{1}$ are allowed we need only use the last four rows in Table 2.4. The resulting table for converting the quotient $0.p_1 p_2 \cdots p_m$ to its two's complement equivalent $z_0.z_1 z_2 \cdots z_m$ is shown below, with the indices changed to match the different indexing used here.

p_j	c_j	z_j	c_{j-1}
1	0	1	0
1	1	0	0
0	0	1	1
0	1	0	1

Show that $.z_1z_2 \cdot z_m = .p_2 \cdots p_m 1$. Also show that based on the first two rows in Table 2.4 $z_0 = 1 - p_1$.

3.12. Find the square root of 0.011111 using the nonrestoring algorithm.

3.6 REFERENCES

[1] J. J. F. CAVANAGH, *Digital computer arithmetic: Design and implementation,* McGraw-Hill, New York, 1984.

[2] Y. CHU, *Computer organization and microprogramming,* Prentice Hall, Englewood Cliffs, NJ, 1972, chap. 5.

[3] M. D. ERCEGOVAC and T. LANG, "On-the-fly conversion of redundant into conventional representations," *IEEE Trans. on Computers, C-36* (July 1987), 895-897.

[4] K. HWANG, *Computer arithmetic: Principles, architecture, and design,* Wiley, New York, 1978.

[5] O. L. MacSORLEY, "High-speed arithmetic in binary computers," *Proc. of IRE, 49* (Jan. 1961), 67-91.

[6] G. W. REITWIESNER, "Binary arithmetic," in *Advances in computers,* vol. 1, F. L. Alt, (Editor), Academic, New York, 1960, pp. 231-308.

[7] J. E. ROBERTSON, "A new class of digital division methods," *IRE Trans. on Electronic Computers, EC-7* (Sept. 1958), 218-222.

[8] N. R. SCOTT, *Computer number systems and arithmetic,* Prentice Hall, Englewood Cliffs, NJ, 1985.

[9] C. TUNG, "Arithmetic," in *Computer science,* A. F. Cardenas et al. (Eds.), Wiley-Interscience, New York, 1972, chap. 3.

[10] S. WASER and M. J. FLYNN, *Introduction to arithmetic for digital system designers,* Holt, Rinehart, Winston, New York, 1982.

4

BINARY FLOATING-
POINT NUMBERS

4.1 PRELIMINARIES

To obtain a dynamic range of representable real numbers without having to scale the operands, we use floating-point numbers instead of fixed-point ones. The representation of floating-point numbers is similar to the commonly used scientific notation and consists of two parts, the *significand* (or mantissa) M and the *exponent* (or characteristic) E. The floating-point number F represented by the pair (M, E) has the value

$$F \;=\; M \;\cdot\; \beta^E$$

where β is the *base* of the exponent. This base is common to all floating-point numbers in a given system. It is therefore not included in the representation of a floating-point number, but is rather implied.

Thus, the n bits that represent a floating-point number are partitioned into two parts, one holding the significand M and the other the exponent E. The range of representable floating-point numbers is larger than that of fixed-point representation, but the precision is smaller. The total number of different values (representable in n bits) is still 2^n, and since the range between the smallest and the largest representable values increases, the distance between any two consecutive values must increase as well. Floating-point numbers are thus sparser than fixed-point numbers, resulting in a lower precision. Any real number whose value lies between two consecutive floating-point numbers is mapped onto one of these two numbers. Therefore, a larger distance between the two consecutive numbers results in a lower precision of representation. A more detailed discussion on the precision of representation appears in Section 4.3.

The significand M and the exponent E of a floating-point number F are both signed quantities. The exponent is usually a signed integer, while the significand is usually represented in one of two ways: one as a pure fraction, and the other as a number in the range $[1, 2)$ (for $\beta = 2$). In addition, different schemes for representing negative values can be employed for each of the two parts of the floating-point number. Until 1980 there was no standard for floating-point numbers and almost every computer system had its own representation method. This made the transportation of scientific programs and data between two different machines very difficult. At that time the IEEE standard 754 [9] was formulated; it is used in most floating-point arithmetic units designed in recent years. The details of this standard are presented in Section 4.4.

Although only very few computer systems still use their own floating-point format rather than the IEEE standard format, it is important to understand some of the prior formats. These prior formats greatly influenced the decisions made by the IEEE floating-point standard committee and studying them allows a better understanding of the IEEE standard format. These formats differ in the partitioning of the n bits between the significand and exponent fields, in the representation method used for each of the two parts, and in the value of the base β. In what follows we will consider only a few prior formats; the others can be analyzed in a similar manner.

We start with the significand field and examine the common case where the significand is a signed-magnitude fraction. The floating-point format in such a case consists of a sign bit S, e bits of an exponent E, and m bits of an unsigned fraction M, satisfying $m + e + 1 = n$, as shown below:

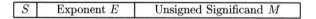

The value of such a floating-point number (S, E, M) is given by

$$F = (-1)^S \cdot M \cdot \beta^E \qquad (4.1)$$

since $(-1)^0 = 1$ and $(-1)^1 = -1$. The maximal value of the fractional significand, denoted by M_{max}, equals $M_{max} = 1 - ulp$, where ulp is the weight of the least-significant bit of the fractional significand. Usually, but, as we will see later, not always, $ulp = 2^{-m}$.

Next we discuss the selection of a value for the implied base β of the exponent. For practical purposes, the base β is restricted to an integer power of the radix $r = 2$. In other words, $\beta = 2^k$ where $k = 1, 2, \cdots$. The reason for this is that it provides a simple method of decreasing the significand and increasing the exponent (and vice versa) at the same time, so that the value of the floating-point number remains unchanged. Whenever an arithmetic operation results in a significand larger than the maximum allowed value of $M_{max} = 1 - ulp$, we must reduce the significand to be in the allowable range. We have to simultaneously increase the exponent so that the value of the floating-point number stays the same. The smallest increase in E, being an integer, is by 1. Therefore, we should use the following relation when the need to reduce the significand arises:

$$M \cdot \beta^E = (M/\beta) \cdot \beta^{E+1}$$

The divide operation in M/β turns into a simple arithmetic shift right operation if β is an integral power of the radix. If $\beta = r = 2$, then shifting the significand to the right by a single position must be compensated by adding 1 to the exponent.

Example 4.1

Suppose that an arithmetic operation yields the result $01.10100 \cdot 2^{100}$, which has a significand larger than M_{max}. We should reduce the significand by shifting it one position to the right and increase the exponent by 1, yielding $0.11010 \cdot 2^{101}$. If $\beta = 2^k$, then changing the exponent by 1 is equivalent to shifting the significand by k positions. Consequently, only k-position shifts are allowed; e.g., for $\beta = 4 = 2^2$, $01.10100 \cdot 4^{010} = 0.01101 \cdot 4^{011}$. $\square$

In general, the representation of numbers in a floating-point form is not unique. For example, $0.11010 \cdot 2^{101} = 0.01101 \cdot 2^{110}$. The same value can also be represented with an exponent of 111, but if the significand field is only 5 bits long, the resulting significand would be 0.00110, so we would lose a significant digit. Out of all possible representations we therefore prefer the one with no leading zeros, allowing us to retain the maximum number of significant digits. We call this the *normalized* form. The normalized form also simplifies the comparison between floating-point numbers. A larger exponent indicates a larger overall number and so the significands have to be compared only for equal exponents. Notice that, since in the case of $\beta = 2^k$, the significand can be shifted only by k (or a multiple of k) positions, the significand is considered normalized if there is a nonzero bit in the first k positions. For example, the normalized form of the number $0.00000110 \cdot 16^{101}$ is $0.01100000 \cdot 16^{100}$, and we could not eliminate the single leading 0.

If the fraction is normalized, the range of the significand is smaller than $[0, 1 - ulp]$. The smallest and largest allowable values are instead

$$M_{min} = \frac{1}{\beta} \qquad \text{and} \qquad M_{max} = 1 - ulp$$

Note that the range of normalized fractions does not include the value zero. Hence, a special representation is needed for zero. A possible representation for zero consists of $M = 0$ and any exponent E. Usually, $E = 0$ is preferred, since the representation of zero in floating-point is then identical to its representation in fixed-point, simplifying the execution of a test for zero instruction.

Finally, we discuss the way exponents are represented. The most common representation is as a biased exponent, according to which

$$E = E^{true} + bias$$

where the *bias* is a constant and E^{true} is the true value of the exponent represented in two's complement. The range for E^{true} using the e bits of the exponent field is

$$-2^{e-1} \leq E^{true} \leq 2^{e-1} - 1.$$

The bias is usually selected as the magnitude of the most negative exponent; i.e., 2^{e-1}, yielding

$$0 \leq E \leq 2^e - 1.$$

In this case, we say that the exponent is represented in the *excess* 2^{e-1} method. The advantage of this scheme is that when comparing two exponents (as is needed in add/subtract operations) we may ignore the sign bits and compare them as if they were unsigned numbers. As a result, if a floating-point format has the S, E, and M components in this order, we may compare the floating-point numbers as though they were binary integers in signed-magnitude representation. Another advantage is that the smallest representable number has the same exponent as zero, namely, 0.

Example 4.2

For $e = 7$ the range of exponents in two's complement representation is $-64 \leq E^{true} \leq 63$ with 1000000 and 0111111 representing the values -64 and 63, respectively. When adding the bias of 64, the true value -64 is represented by 0000000 and the true value 63 is represented by 1111111. This representation is called the excess 64 scheme.   □

Note that the excess 2^{e-1} representation can be obtained by simply inverting the sign bit of the two's complement representation; i.e., letting the values 0 and 1 of the sign bit indicate negative and positive numbers, respectively.

The complete range of normalized floating-point numbers includes identical subranges for positive floating-point numbers, denoted by F^+, and negative floating-point numbers, denoted by F^-. The range of positive floating-point numbers is

$$M_{min} \cdot \beta^{Emin} \leq F^+ \leq M_{max} \cdot \beta^{Emax}$$

where E_{min} is the smallest exponent, and E_{max} is the largest. An identical range exists for negative numbers. Whenever the exponent of the result of some arithmetic operation is larger than E_{max}, an *exponent overflow* indication should be generated. Similarly, an exponent smaller than E_{min} should generate an *exponent underflow* indication. These flags make the programmer aware of the situation and allow him/her to take the necessary steps. Since the significand is always kept normalized, any overflow will be reflected through the exponent. In some machines, when an exponent overflow occurs, a special representation of infinity is used for the result. Two other possibilities are stopping the computation and interrupting the processor, or, the least recommended, setting the result to the largest representable number. If an exponent underflow occurs, the representation of zero is used for the result in some cases, but the proper exponent underflow flag is still raised. Setting the result to zero allows the computations to proceed, if appropriate, without any interruptions. A detailed discussion of the way such exceptions are handled in the IEEE standard for floating-point numbers appears in Section 4.8.

The complete range of floating-point numbers is shown in the following schematic diagram. Notice that zero is not included in the range of either F^+ or F^-.

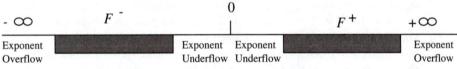

Example 4.3

The short floating-point format in the IBM 370 system consists of 32 bits partitioned as follows:

S – sign bit	E – 7 bits, excess 64 exponent	M – 24 bits, unsigned fractional significand

The base of these floating-point numbers is $\beta = 16$, and hence

$$F = (-1)^S \cdot M \cdot 16^{E-64}.$$

Here, E_{min} is represented by 0000000 and has the value -64, while E_{max}, represented by 1111111, has the value $+63$. Since $\beta = 16$, it is convenient to consider the significand as consisting of six hexadecimal digits. The normalized significand therefore satisfies

$$M_{min} = 16^{-1} \leq M \leq M_{max} = 1 - 16^{-6} = 1 - 2^{-24}.$$

Consequently, $F^+_{max} = (1 - 16^{-6}) \cdot 16^{63} \approx 7.23 \cdot 10^{75}$, and $F^+_{min} = (16^{-1}) \cdot 16^{-64} \approx 5.4 \cdot 10^{-79}$.

	IBM/370	DEC/VAX	Cyber 70
Word length (double)	32 (64) bits	32 (64) bits	60 bits
Significand+{hidden bit}	24 (56) bits	$23 + 1\ (55 + 1)$ bits	48 bits
Exponent	7 bits	8 bits	11 bits
Bias	64	128	1024
Base	16	2	2
Range of M	$\frac{1}{16} \leq M < 1$	$\frac{1}{2} \leq M < 1$	$1 \leq M < 2$
Representation of M	Signed-magnitude	Signed-magnitude	One's complement
Approximate range	$16^{63} \approx 7 \cdot 10^{75}$	$2^{127} \approx 1.9 \cdot 10^{38}$	$2^{1023} \approx 10^{307}$
Approximate resolution	$2^{-24} \approx 10^{-7}\ (10^{-17})$	$2^{-24} \approx 10^{-7}\ (10^{-17})$	$2^{-48} \approx 10^{-14}$

TABLE 4.1   The floating-point formats of three machines.

For example, let $(S, E, M) = (C1200000)_{16}$ be a floating-point number in the short IBM format, then the first byte consists of $(11000001)_2$. The sign bit is $S = 1$; i.e, the number is negative. The exponent is 41_{16} and, with a bias of $64_{10} = 40_{16}$, E^{true} is $(41 - 40)_{16} = 1$. Finally, $M = (0.2)_{16}$, hence $F = (-0.0010)_2 \cdot 16^1 = (-2)_{10}$.

The resolution of this floating-point representation, defined as the distance between two consecutive significands, is equal to the weight of the least-significant bit of the significand. Thus, the resolution of this representation is $ulp = 16^{-6} = 2^{-24} \approx 0.6 \cdot 10^{-7}$. We say that the short format has approximately seven significant decimal digits. Should a higher precision be desired, the IBM system provides a long floating-point format, in which the significand is extended by adding to it a second 32-bit word. This format is

sign bit	7 bits - excess 64 exponent	56 bits - unsigned fractional significand

The range is roughly the same, but the resolution is now $ulp = 16^{-14} = 2^{-56} \approx 10^{-17}$; i.e., 17, instead of 7, significant decimal digits.   □

Table 4.1 compares the floating-point format in IBM computers to those used in DEC/VAX and CDC/Cyber 70 computers. The hidden bit in the DEC/VAX is a scheme to increase the number of significant bits in the significand and thus increase the precision. For a base β of 2 the normalized significand will always have a leading 1. This bit can be eliminated, allowing the inclusion of an extra bit. As a result, the resolution becomes $ulp = 2^{-24}$ instead of 2^{-23}. The value of a floating-point number (S, f, E) in the short DEC format is therefore

$$(-1)^S 0.1f \cdot 2^{E-128}$$

where f is the pattern of 23 bits in the significand field. In this case, a zero significand field ($f = 0$) represents the fraction $0.10_2 = 1/2$. Consider now the case where $f = 0$ and $E = 0$. With a hidden bit, this may represent the value $0.1 \cdot 2^{0-128} = 2^{-129}$. However, the floating-point number $f = E = 0$ is still expected to represent 0, a representation which does not use a hidden bit. We clearly can not allow $f = E = 0$ to represent two different values. To avoid this from happening we must restrict the use of the exponent $E = 0$ and reserve it for representing the value zero only. Consequently, the smallest exponent allowed for nonzero numbers is $E = 1$. Therefore, the smallest positive number in the DEC/VAX system is $F^+_{min} = \frac{1}{2}2^{1-128} = 2^{-128}$. The largest positive number is $F^+_{max} = (1 - 2^{-24}) \cdot 2^{255-128} = (1 - 2^{-24}) \cdot 2^{127}$.

4.2 FLOATING-POINT OPERATIONS

The way floating-point operations are executed depends on the specific format used for representing the operands. In what follows we will assume that the significands are normalized fractions in signed-magnitude representation and that the exponents are biased. Given two numbers, $F_1 = (-1)^{S_1} \cdot M_1 \cdot \beta^{E_1-bias}$ and $F_2 = (-1)^{S_2} \cdot M_2 \cdot \beta^{E_2-bias}$, we need to calculate the result of a basic arithmetic operation yielding $F_3 = (-1)^{S_3} \cdot M_3 \cdot \beta^{E_3-bias}$. We start with multiplication and division, since these are easier to follow than addition and subtraction, which will be described later on.

   Multiplication.   The significands of the two operands are to be multiplied as if they were fixed-point numbers. The exponents of the operands are to be added. These two operations can be done in parallel. The sign S_3 is positive if the signs S_1 and S_2 are equal and is negative otherwise. When adding the two exponents $E_1 = E_1^{true} + bias$ and $E_2 = E_2^{true} + bias$, the bias should be subtracted once to obtain the correct exponent. For $bias = 2^{e-1}$ (which in binary is represented as $100\cdots0$), subtracting the bias is equivalent to adding the bias and is accomplished by complementing the sign bit. If the resulting exponent E_3 is larger than E_{max}, an overflow indication must be generated. If the exponent E_3 is negative and is smaller than E_{min}, then an underflow indication must be generated.

   When multiplying the significands we have to make sure that M_3 is a normalized significand. Since each operand's significand satisfies $1/\beta \le M_i < 1$, $(i = 1, 2)$, the product of the two significands satisfies

$$1/\beta^2 \le M_1 \cdot M_2 < 1.$$

Consequently, we may need to shift the significand one position to the left in order to normalize it. This is achieved by performing one base-β left shift operation; i.e., k base-2 shifts for $\beta = 2^k$, at the same time reducing the exponent by 1. This is called the *postnormalization* step. After this step is executed, the exponent may become smaller than E_{min}, and an exponent underflow indication should be generated.

Division.   The significands of the two operands are to be divided and the exponents subtracted. Here, we have to add the bias to the difference $E_1 - E_2$. If the resulting exponent is out of the range, an overflow or underflow indication should be generated. The resultant significand satisfies

$$1/\beta \leq M_1/M_2 < \beta.$$

Therefore, a single base-β shift right of the significand, accompanied by an increase of one in the exponent, may be required in the postnormalization step. The exponent increase may, in turn, lead to an overflow.

If the divisor is zero, an overflow occurs, and a special indication of *division by zero* should be generated and the quotient can be set to $\pm\infty$. If both divisor and dividend are zero, the result is undefined and in the IEEE 754 standard such a quantity has a special representation called *not a number* (NaN). NaN also represents uninitialized variables and the result of $0 \cdot \infty$. These will be discussed further in Section 4.4.

Remainder.   Unlike fixed-point division, floating-point division does not generate a final remainder. The fixed-point remainder, denoted by R, is defined as $X - QD$ where X, Q, and D are the dividend, quotient and divisor, respectively (see Chapter 3). This remainder satisfies the inequality $|R| \leq |D|$ and is a byproduct of the direct division algorithms, like the restoring and nonrestoring algorithms. The situation is different in floating-point division. The floating-point remainder, denoted by $F_1 \ REM \ F_2$, is defined as $F_1 - F_2 \cdot Int(F_1/F_2)$, where $Int(F_1/F_2)$ is the quotient F_1/F_2 converted to an integer. The conversion of the quotient to an integer can be performed either through truncation (i.e., removing the fractional part) or through rounding-to-nearest. The IEEE standard uses the round-to-nearest-even mode which is defined in Section 4.5. In this case, the following inequality is satisfied:

$$|F_1 \ REM \ F_2| \leq |F_2|/2$$

Careful examination of the expression for $F_1 \ REM \ F_2$ reveals the higher complexity involved in calculating the floating-point remainder compared to that of the fixed-point remainder. An algorithm for floating-point division will generate a quotient represented as a floating-point number and will not generate the integer $Int(F_1/F_2)$ which can be as large as $\beta^{Emax-Emin}$. Therefore, we must calculate the floating-point remainder separately so that we can perform this time-consuming calculation only when it is required. The floating-point remainder is needed, for example, when performing an argument reduction for periodic functions like the trigonometric functions sine and cosine.

A brute-force method would be to continue the execution of a direct division algorithm employed for calculating M_1/M_2 for $E_1 - E_2$ steps, even if $E_1 - E_2$ is much greater than the number of steps needed to generate the m bits of the quotient's significand. In practice, this is not an acceptable solution since the execution of the floating-point remainder operation may take an arbitrary number of clock cycles. As a result, the floating-point remainder is often calculated in software rather than in hardware. An alternative solution is to define a *REM*-step operation, X *REM* F_2, which performs a pre-specified maximum number of divide steps, such as the number of divide steps required in a regular divide operation. Initially, X is equal to F_1, afterwards it is made equal to the remainder of the previous *REM*-step operation. Such a *REM*-step operation can be repeated until a remainder that is smaller than $F_2/2$ is obtained [7].

Addition/Subtraction.   These operations require that the exponents of both operands be equal before adding or subtracting the significands. Only when $E_1 = E_2$ can the term β^{E_1} be factored out and the two significands M_1 and M_2 be added. To achieve this we *align* the significands by shifting the significand of the smaller operand to the right, increasing its exponent at the same time, until it equals the other exponent. In other words, the significand of the smaller number (i.e., the number with the smaller exponent) is shifted $|E_1 - E_2|$ base-β positions to the right. For example, if $E_1 \geq E_2$, then

$$F_1 \pm F_2 = \left((-1)^{S_1} \cdot M_1 \pm (-1)^{S_2} \cdot M_2 \cdot \beta^{-(E_1-E_2)}\right) \cdot \beta^{E_1-bias}. \qquad (4.2)$$

Note that we do not decrease the exponent of the larger number to make it equal to the other exponent, since this will result in a significand larger than 1, and a larger significand adder will be required.

If, based on the two sign bits and the originally required operation, an addition is performed, then the resultant significand denoted by M (obtained by adding the two aligned significands), is in the range

$$1/\beta \leq M < 2.$$

If the significand M is greater than 1, a postnormalization step is required. This consists of shifting the significand to the right to yield M_3, and increasing the exponent by one. At this point an exponent overflow may occur. In summary, the following steps are required when adding or subtracting two floating-point numbers:

Step 1: Calculate the difference d of the two exponents, $d = |E_1 - E_2|$.

Step 2: Shift the significand of the smaller number by d base-β positions to the right.

Step 3: Add the aligned significands and set the exponent of the result equal to the exponent of the larger operand.

Step 4: Normalize the resultant significand and adjust the exponent if necessary.

Step 5: Round the resultant significand and adjust the exponent if necessary.

If the final operation called for is subtraction, then the resultant significand satisfies

$$0 \le |M| < 1$$

and a postnormalization step is required if the resultant significand is smaller than $1/\beta$. This step consists of shifting the significand to the left and decreasing the exponent simultaneously, which may lead to an exponent underflow. In extreme cases, the postnormalization step may require a shift left operation over all bits in the significand, yielding a zero result.

Example 4.4

Let $F_1 = (0.100000)_{16} \cdot 16^3$ and $F_2 = (0.FFFFFF)_{16} \cdot 16^2$ be two numbers in the short IBM format, to be subtracted. The significand of the smaller one (i.e., F_2), has to be shifted to the right, resulting in the loss of the least-significant digit:

	0.	1	0	0	0	0	0	·	
F_1	0.	1	0	0	0	0	0	·	16^3
F_2 aligned	0.	0	F	F	F	F	F	·	16^3
$F_1 - F_2$	0.	0	0	0	0	0	1	·	16^3
Postnormalization	0.	1	0	0	0	0	0	·	16^{-2}

Not only is this a time-consuming postnormalization step (shifting over five hexadecimal digits), but the final result is in error. The correct result (with an "unlimited" number of significand digits) is

	0.	1	0	0	0	0	0	·	
F_1	0.	1	0	0	0	0	0	·	16^3
F_2 aligned	0.	0	F	F	F	F	F	·	16^3
$F_1 - F_2$	0.	0	0	0	0	0	1	·	16^3
Postnormalization	0.	1	0	0	0	0	0	·	16^{-3}

The error (also called loss of significance) is $0.1 \cdot 16^{-2} - 0.1 \cdot 16^{-3} = 0.F \cdot 16^{-3}$. A solution to this problem would be to have guard digits; i.e., additional digits to the right of the significand to hold the shifted-out digits. In the above example, a single (hexadecimal) guard digit is sufficient. These guard digits will be discussed in Section 4.5.   □

A simplified block diagram of the circuitry required to perform the addition or subtraction of floating-point numbers is depicted in Figure 4.1. The

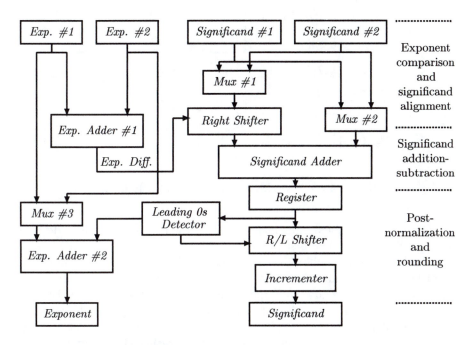

FIGURE 4.1   Floating-point adder/subtractor.

block diagram shows the two separate data paths, with the left one for the exponents and the right one for significands. *Exp. Adder #1* computes the difference between the exponents of the two operands. The resulting exponent difference (*Exp. Diff.* in the figure) determines the amount of shift right positions that the significand of the smaller operand must go through in order to be aligned with the other significand. *Mux #1* is a multiplexor (also known as data selector) which routes one of its two inputs to its single output. The selection of the input significand to be routed to the *Right Shifter* is determined by the sign of the exponent difference. This sign also controls *Mux #2*, which selects the significand of the larger operand and routes it to the *Significand Adder*. In the next step the addition or subtraction of the now aligned significands is performed and the result stored in a register. Then, a special circuit, *Leading 0s Detector*, examines the leading bits of the resulting significand and determines the type of shift operation needed for the postnormalization step and the corresponding adjustment of the exponent. This adjustment is performed using *Exp. Adder #2* whose second input is the exponent of the larger operand. This, in turn, is selected by *Mux #3* which is again controlled by the sign of the exponent difference. Finally, the *Incrementer* executes the rounding, if necessary, according to the rules explained in the next section.

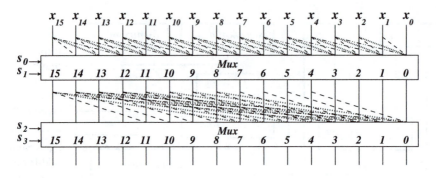

FIGURE 4.2   A two-level radix-4 shifter for 16 bits.

Out of the functional units included in Figure 4.1 the two shifters deserve a separate discussion. The first shifter should be capable of performing right (alignment) shifts only while the second one should be able to perform either right or left (postnormalization) shifts. More importantly, the two shifters must be capable of performing large shift operations, as large as the number of digits in the significand field. The overall performance of the floating-point add/subtract unit is highly dependent on the speed of these two shifters. Consequently, they are usually implemented as combinatorial shifters rather than shift registers, which would require a large and variable number of clock cycles to complete the shift. A combinatorial shifter generates all possible shifted patterns but only one is provided at the output according to some control bits. Since, in general, such combinatorial shifters are capable of performing circular shifts (rotates) as well, they are commonly known as *barrel shifters*.

A barrel shifter can be implemented as a single level array where each input bit is directly connected to m (and even more) output lines. For $m = 53$ (the number of significand bits in the IEEE double-precision format, see Section 4.4) the large number of connections (and the resulting large electrical load) make this an undesirable solution, although the overall design is conceptually simple [14]. One alternative is a two-level array. We can implement a two-level combinatorial shifter for 53 bits by having the first level shift the bits by 0, 1, 2 or 3 bit positions, and letting the second level shift the bits by multiples of 4 (i.e., 0, 4, 8, $\cdots$, 52). In this way, shifts from length 0 to 53 can be performed. We call this two-level shifter a radix-4 shifter. An example of a two-level radix-4 shifter for 16 bits is shown in Figure 4.2. In the first level of the 53-bit radix-4 shifter each bit has four destinations, while in the second level each has 14 destinations. A more balanced two-level shifter for 53 bits would be the radix-8 shifter, where the first level shifts from 0 to 7 bit positions and the second level shifts by multiples of 8 (i.e., 0, 8, 16, 24, 32, 40 and 48). Thus, each bit in the first level has 8 destinations and 7 in the second level.

4.3 CHOICE OF FLOATING-POINT REPRESENTATION

Although the IEEE standard 754 described in the next section is now commonly used in newly designed arithmetic units, it is important to understand the implications of selecting a particular format and, as a result, understand the reasons behind the adopted standard.

When designing a format for floating-point numbers we are given n, the total number of bits, and we have to determine the length of the significand and exponent fields, denoted by m and e, respectively (satisfying $m + e + 1 = n$), and the value of the exponent base β. One goal that we would like to achieve is having a small representation error, which is the error made when representing a high-precision real number in a finite-length floating-point format. Let x be a real number and $Fl(x)$ be its machine representation. $Fl(x) - x$ is called the *absolute representation error*. For every real number x that is within the range of the floating-point numbers, there are two consecutive representations F_1 and F_2 satisfying $F_1 \leq x \leq F_2$. Thus, we have the choice of setting $Fl(x)$ equal to either F_1 or F_2. If $F_1 = M\beta^E$, then $F_2 = (M + ulp)\beta^E$, and the maximum absolute error is half the distance between F_1 and F_2, which in turn equals $ulp \cdot \beta^E$. Unlike the distance between two consecutive significands (ulp), which was discussed in the previous section, the distance between two consecutive floating-point numbers $(ulp \cdot \beta^E)$ is clearly not a constant but varies, becoming larger as the exponent increases, as shown in the following diagram:

Usually, the absolute size of the error is less important than its relative size (compared to the original value x). Thus, a more important and commonly used measure for the representation error is $\delta(x) = (Fl(x) - x)/x$, which is the *relative representation error*. To measure the "accuracy" of the representation we may use the maximum relative representation error (MRRE), which is an upper bound of $\delta(x)$. This upper bound can be obtained as follows:

$$\delta(x) \ \leq \ \frac{\frac{1}{2}\,ulp\,\beta^E}{M\beta^E} \ = \ \frac{1}{2}\,\frac{ulp}{M} \ \leq \ \frac{1}{2}\,\frac{ulp}{\frac{1}{\beta}} \ = \ \frac{1}{2}\,ulp \cdot \beta \qquad (4.3)$$

Thus, the MRRE $= \frac{1}{2}\,ulp \cdot \beta$ increases with the exponent base β but decreases with ulp; that is, with the number of significand bits, m.

The MRRE would provide an acceptable measure for the accuracy of the representation if the operands in floating-point computations were more or less uniformly distributed. However, as has been observed, the distribution of floating-point operands is not uniform but approaches the reciprocal distribution with the following density function [20]:

$$\frac{1}{M \ln \beta} \; ; \qquad 1/\beta \leq M \leq 1$$

In other words, larger significands are less likely to occur than smaller significands. For example, the first digit of a decimal floating-point operand will most likely be a 1; 2 is the second most likely and so on.

This nonuniform distribution is taken into account in the second measure proposed for representation error. This measure is the *average* relative representation error (ARRE). The maximum value of the absolute error is, as has been shown above, $\frac{1}{2} ulp \cdot \beta^E$, and since the minimum error is zero, the average absolute error is $\frac{1}{4} ulp \, \beta^E$. The corresponding relative representation error is therefore $\frac{1}{4} \frac{ulp}{M}$ and, consequently,

$$\text{ARRE} = \int_{\frac{1}{\beta}}^{1} \frac{1}{M \ln \beta} \frac{ulp}{4M} \, dM \; = \; \frac{\beta - 1}{\ln \beta} \frac{ulp}{4}. \qquad (4.4)$$

Another factor to be considered when selecting a floating-point format is the size of the range. The range of the positive floating-point numbers, for example, is approximately equal to the largest positive number representable, since the smallest positive number is very close to zero. We will use therefore the expression $\beta^{E_{max}}$ for the range. Thus, to obtain a large range we should increase β and/or the number of exponent bits e. Increasing the latter implies less bits for the significand field within the floating-point format and a higher value of ulp, resulting in a higher representation error. A similar effect is experienced if, instead of increasing e, we increase the exponent base β. Consequently, there is a tradeoff between the range and the representation error.

We may consider several floating-point representations that have the same range and select the one with the smallest MRRE or ARRE. Another possibility is to consider several representations with the same MRRE (or ARRE) and select the one with the largest range. For example, for a 32-bit word $m + e$ is 31 (one bit is reserved for the sign), and we may use one of the representations shown in Table 4.2 [4]. All three have nearly the same range. The base 16 representation is inferior to the other two representations with $\beta = 2$ and $\beta = 4$ if the MRRE is selected as a measure for the representation error. Base 4 turns out to produce the lowest ARRE for this particular example. If base 2 is selected, and a hidden bit is used, then the MRRE and ARRE are reduced by a factor of 2, making this format the one with the smallest representation error.

A different goal that one might consider when deciding upon a floating-point format is the execution time of floating-point operations. Two time-

β	e	m	Range	MRRE	ARRE
2	9	22	$2^{2^8-1} = 2^{255}$	$0.5 \cdot 2^{-22} \cdot 2 = 2^{-22}$	$0.180 \cdot 2^{-21}$
4	8	23	$4^{2^7-1} = 2^{2^8-2} = 2^{254}$	$0.5 \cdot 2^{-23} \cdot 4 = 2^{-22}$	$0.135 \cdot 2^{-21}$
16	7	24	$16^{2^6-1} = 2^{2^8-4} = 2^{252}$	$0.5 \cdot 2^{-24} \cdot 16 = 2^{-21}$	$0.169 \cdot 2^{-21}$

TABLE 4.2 Range, MRRE, and ARRE of three 32-bit floating-point formats (4).

consuming steps are the aligning of the significands before add/subtract operations and postnormalization in any floating-point operation. It has been observed that a larger exponent base β yields a higher probability of equal exponents in add/subtract operations (in which case no alignment step is necessary) and a lower probability that a postnormalization step will be needed. Statistical analysis of a variety of programs has provided the results included in Table 4.3 [19]. This table shows the percentage of cases in which no alignment shifts were needed, those in which a single alignment shift was required, and those in which a larger (two or more positions) shift was needed. Also, postnormalization was not required in 59.4% of the cases for $\beta = 2$, while it was not needed in 82.4% of the cases for $\beta = 16$. It is interesting to note that even for $\beta = 2$, no postnormalization step is necessary in most cases. The above consideration is of limited practical significance when a barrel shifter, as described in Section 4.2, is used.

Alignment shift	$\beta = 16$	$\beta = 2$
0	47.3%	32.6%
1	26.0%	12.1%
≥ 2	26.7%	55.3%

TABLE 4.3 The probability of alignment shifts of different sizes (19).

4.4 THE IEEE FLOATING-POINT STANDARD

The IEEE floating-point standard defines four formats for floating-point numbers. The first two are the basic single-precision 32-bit format and the double-precision 64-bit format. The other two are the extended formats, to be used for intermediate results. The single extended format should have at least 44 bits, and the double extended format should have at least 80 bits. These extended formats have a higher precision and a higher range than the corresponding 32- and 64-bit formats.

4.4.1 Single-Precision Format

The most important objective for the 32-bit format is precision of representation. Hence, base 2 was selected, allowing the use of a hidden bit to further increase the precision. As a result, the suggested format is similar to the DEC format (see Table 4.1), but there are some differences, as indicated below. In order to have a reasonable range, an exponent field of length 8 bits was selected, yielding the following format:

S	8 bits - biased exponent E	23 bits - unsigned fraction f

Out of the 256 combinations of the exponent field, two are reserved for special values. $E = 0$ is reserved for zero (with fraction $f = 0$) and denormalized numbers (with fraction $f \neq 0$). $E = 255$ is reserved for $\pm\infty$ (with fraction $f = 0$) and NaN, (with fraction $f \neq 0$). These special representations are further discussed below. For the remaining exponents (i.e., $1 \leq E \leq 254$), the value of the floating-point number is given by

$$F = (-1)^S \; 1.f \; 2^{E-127}. \tag{4.5}$$

There are two differences between the IEEE single precision format and the DEC short format. The exponent bias is 127 instead of $2^{e-1} = 2^7 = 128$. This provides a larger maximum value of the true exponent, $254 - 127 = 127$ instead of $254 - 128 = 126$, yielding a larger range. A similar effect is achieved by using a significand of $1.f$ instead of $0.f$, since this adds 1 to the exponent. As a result, the largest and smallest positive numbers are

$$F_{max}^+ = (2 - 2^{-23}) \cdot 2^{254-127} = (1 - 2^{-24}) \cdot 2^{128}$$

and $$F_{min}^+ = 1.0 \cdot 2^{1-127} = 2^{-126}$$

compared to $F_{max}^+ = (1 - 2^{-24}) \cdot 2^{127}$ and $F_{min}^+ = 2^{-128}$, respectively, in the DEC format. The exponent bias and significand range were selected so as to allow the reciprocal of all normalized numbers (in particular, F_{min}^+) to be represented without overflow. This requirement is not satisfied for F_{min}^+ in the DEC format.

Finally, a few comments about the special values that can be represented in the IEEE format and which are summarized in the following table.

	$f = 0$	$f \neq 0$
$E = 0$	0	Denormalized
$E = 255$	$\pm\infty$	NaN

Operations dealing with the values $\pm\infty$ that are represented by $f = 0$, $E = 255$, and $S = 0, 1$ must obey the traditional mathematical conventions such as $F + \infty = \infty$, $F/\infty = 0$, etc. The denormalized numbers provide representations for values smaller than the smallest normalized number, lowering the probability of an exponent underflow. Denormalized numbers are represented by $E = 0$, and their value is given by

$$F = (-1)^S \; 0.f \; 2^{-126}. \tag{4.6}$$

This is sometimes expressed as $(-1)^S\ 0.f\ 2^{1-127}$ to have the same bias as normalized numbers. Note that denormalized numbers have no hidden bit since the significands should not be normalized. Also, although the true value of the exponent should have been $0 - 127 = -127$, the value -126 was selected, since the smallest normalized number is $F^+_{min} = 1 \cdot 2^{-126}$. With denormalized numbers the smallest representable number is $2^{-23} \cdot 2^{-126} = 2^{-149}$ instead of 2^{-126}. The addition of denormalized numbers has been termed gradual underflow or graceful underflow. It does not eliminate underflow, but it substantially reduces the gap between the smallest representable number and zero. This gap, of size 2^{-149}, is equal to the distance between any two consecutive denormalized numbers and is also the distance between any two consecutive normalized numbers with the smallest possible exponent $(1 - 127) = -126$, as illustrated in the following diagram:

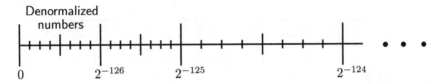

Denormalized numbers have not been included in all the designs of arithmetic units that follow the IEEE standard. This is mainly due to the high cost associated with their implementation, since the representation of denormalized numbers is different from that of normalized numbers, requiring a more complex design and possibly a longer overall execution time. Even designs that implement denormalized numbers allow the programmer to avoid their use if faster execution is desired.

The IEEE standard also defines a single-extended format to be employed when calculating intermediate results within the evaluation of complex functions like the transcendental and power function. The single-extended format extends the exponent field from 8 to 11 bits and the significand field from 23+1 to 32 or more bits (without a hidden bit). Thus, the total length of a single-extended floating-point number is at least 1+11+32=44 bits.

There are two kinds of NaN (Not a Number), the signaling (or trapping) NaN, and the quiet (or nontrapping) NaN. NaNs are represented in the single-precision format by E=255 and $f \neq 0$ allowing a large number of possible values. The most significant bits of the fraction can be used to distinguish between the two kinds of NaNs. The remaining bits may contain system-dependent information. An example of a signaling NaN is an uninitialized variable. A signaling NaN sets the Invalid operation exception flag (see Section 4.8) whenever any arithmetic operation with this NaN as an operand, is attempted. In contrast, a quiet NaN does not set the Invalid operation exception flag when involved in an arithmetic operation. A signaling NaN turns into a quiet NaN when used as an operand for an arithmetic operation if the Invalid operation trap is disabled, to avoid setting the Invalid operation flag again later on. A quiet NaN is also produced when an invalid operation such as $0 \cdot \infty$ is attempted, since this operation had already set the Invalid operation flag once. The fraction field in a quiet NaN may contain a pointer to the offending line of code. A quiet NaN, when used as an operand of an arithmetic operation will produce the same quiet NaN as a result and will not set any exception flag. For example, NaN+5=NaN. If both operands of an arithmetic operation are quiet NaNs, the result will equal the NaN with the smallest significand.

4.4.2 Double-Precision Format

The main consideration for the double-precision format is range. Consequently, the exponent field is increased to 11 bits yielding the following format:

S	11 bits - biased exponent E	52 bits - unsigned fraction f

The extreme values of E, i.e., 0 and 2047, are reserved for the same purposes as in the single-precision format. The value of a floating-point number with an exponent E in the range $1 \le E \le 2046$ is

$$F \; = \; (-1)^S \; 1.f \; 2^{E-1023}. \tag{4.7}$$

This format, as well as the single-precision format, are summarized in Table 4.4. A double-extended format is also defined in the IEEE standard. It extends the exponent field from 11 to 15 bits and the significand field from 52+1 to 64 or more bits (i.e., without a hidden bit), and consequently the total number of bits in the double-extended floating-point format is at least 1+15+64=80. The interested reader is referred to several articles describing the details of the IEEE floating-point standard that appear in the *IEEE Computer Magazine,* Vol. 14, March 1981.

	Single	Double
Word length	32 bits	64 bits
Fraction + hidden bit	23 + 1 bits	52 + 1 bits
Exponent	8 bits	11 bits
Bias	127	1023
Approximate range	$2^{128} \approx 3.8 \cdot 10^{38}$	$2^{1024} \approx 9 \cdot 10^{307}$
Smallest normalized number	$2^{-126} \approx 10^{-38}$	$2^{-1022} \approx 10^{-308}$
Approximate resolution	$2^{-23} \approx 10^{-7}$	$2^{-52} \approx 10^{-15}$

TABLE 4.4   The single and double IEEE floating-point formats.

4.5 ROUND-OFF SCHEMES

The accuracy of results obtained in a floating-point arithmetic unit is limited even if the intermediate results calculated in the arithmetic unit are accurate. The number of computed digits may exceed the total number of digits allowed by the format, and we have to dispose of the extra digits before the final results are stored in a user-accessible register or in the memory. For example, when multiplying two significands each of length m, a product of length $2m$ is generated and we must round it off to m digits.

When selecting a round-off scheme we need to consider the following:

1. Accuracy of results (numerical considerations).

2. Cost of implementation and speed (machine considerations).

Let x and y be real numbers and let Fl be the set of machine representations in a given floating-point format. Denote by $Fl(x)$ the machine representation of x. When rounding real numbers to machine representations the following conditions should be satisfied:

1. $Fl(x) \leq Fl(y)$ whenever $x \leq y$.

2. If $x \in Fl$ then $Fl(x) = x$.

3. If F_1 and F_2 are two consecutive numbers in Fl such that $F_1 \leq x \leq F_2$, then either $Fl(x) = F_1$ or $Fl(x) = F_2$.

Let d denote the number of extra digits that are kept in the arithmetic unit (in addition to the m significand digits) before rounding is performed. For convenience, let us assume that there is a radix point between the m most significant digits (of the significand) and the d extra digits. Thus, we will investigate ways to round numbers like 2.99_{10} and obtain an integer.

The simplest scheme is called *truncation* or *chopping* and is illustrated in Figure 4.3 [11]. We remove the d extra digits with no change in the m remaining digits. For a given $F_1 \leq x \leq F_2$, $Trunc(x)$ results in rounding toward zero, yielding the smaller of F_1 and F_2. For example, the decimal number $x = 2.99$, when rounded toward zero, yields 2. This is a fast method that does not require any extra hardware, but its numerical performance is very poor. The error introduced by truncation can be almost as large as ulp (the weight of the least-significant digit of the significand).

The curve for $Trunc(x)$ lies entirely below the ideal line (the dotted line in Figure 4.3) which provides infinite precision. We say that truncation has a *negative bias* where the bias, in general, measures the tendency of a round-off scheme to favor errors of a particular sign. Clearly, we would like to use a round-off scheme that is unbiased, or has a very small bias. To compare the bias of truncation to that of other rounding schemes quantitatively, we define

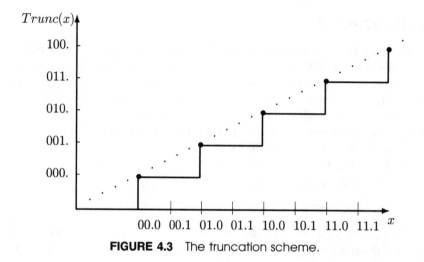

FIGURE 4.3 The truncation scheme.

the bias for a given d as the average error for a set of 2^d consecutive numbers, where Error $= Trunc(x) - x$ and a uniform distribution for the significand is assumed [11].

For example, the rounding errors when truncating with $d = 2$ are shown in Table 4.5. In this table, X is any significand of length m. The sum of errors for all $2^d = 4$ consecutive numbers is $-3/2$. Therefore, the bias for $d=2$, which is the average error, equals $-3/8$.

A more accurate scheme is the *round-to-nearest* scheme (commonly known as just *rounding*), which for $F_1 \le x \le F_2$ yields the nearer to x out of F_1 and F_2. It is obtained by adding 0.1_2 to x (or in general, adding half a *ulp*) and retaining only the integer part of the sum (chopping the fraction). For example, to round off the decimal number $x = 2.99$, we add 0.5 and chop off the fractional part of 3.49, obtaining 3. The maximum error would occur when $x = 2.5$. Rounding it off yields $2.50 + 0.50 = 3.00$ finally resulting in 3, with an error of 0.5. Round-to-nearest is used in many arithmetic units and its curve is shown in Figure 4.4 [11]. Note that for performing the round-to-nearest scheme, a single extra digit (i.e., $d=1$) is sufficient.

Number	$Trunc(x)$	Error
$X.00$	X	0
$X.01$	X	$-1/4$
$X.10$	X	$-1/2$
$X.11$	X	$-3/4$

TABLE 4.5 The rounding errors for the truncation scheme with $d = 2$.

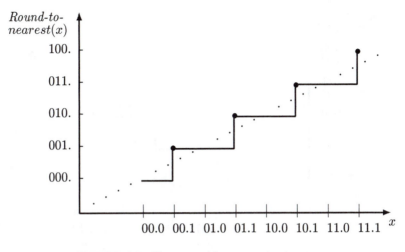

FIGURE 4.4   The round-to-nearest scheme.

Round(x) is nearly symmetric with respect to the ideal line, which is a substantial improvement over truncation. However, at the point $X.10$ we would always round up (this is indicated in Figure 4.4 by the heavy dots), and over a long sequence of operations we may get a slight positive bias. For $d = 2$ this bias can be calculated from Table 4.6 [11]. The sum of errors for all $2^d = 4$ numbers is $+1/2$, and thus the bias is $1/8$, which is smaller than the bias of the truncation scheme. The same sum of errors (i.e., $+1/2$) is obtained for $d > 2$ and consequently the bias, in general, is $\frac{1}{2} \cdot 2^{-d}$. This indicates that the positive bias is due to the rounding up of $X.10\cdots0$. To obtain an unbiased rounding we could, in case of a tie (i.e., $X.10$), choose out of F_1 and F_2, the one whose least-significant bit is 0 (i.e., the even one). This way we would alternately round up and down. The obtained scheme is called the *round-to-nearest-even* scheme and is illustrated in Figure 4.5 [11].

Number	*Round-to-nearest(x)*	Error
$X.00$	X	0
$X.01$	X	$-1/4$
$X.10$	$X + 1$	$+1/2$
$X.11$	$X + 1$	$+1/4$

TABLE 4.6   The rounding errors for the round-to-nearest scheme with $d = 2$.

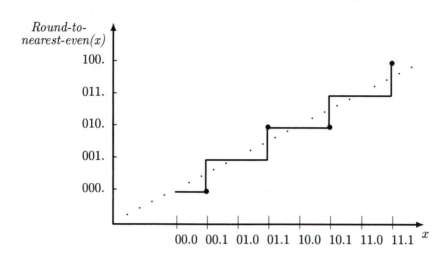

FIGURE 4.5   The round-to-nearest-even scheme.

To compute the bias of this scheme for $d = 2$ we have to consider two groups of size $2^d = 4$ as shown in Table 4.7. The sum of errors is $-1/2$ for the first group and $+1/2$ for the second. Thus, the average bias is 0. The round-to-nearest-even scheme is mandatory in the IEEE floating-point standard. Another possible modification to the round-to-nearest scheme, which also yields an unbiased rounding, is, in case of a tie, to choose out of F_1 and F_2 the one whose least-significant bit is 1. This is known as the *round-to-nearest-odd scheme*.

Although the round-to-nearest schemes have a good numerical "performance," their main disadvantage is that they require a complete add operation, since the carry from the least-significant bit may propagate across the entire significand. To avoid this time-consuming carry-propagation, it has been suggested [10] to use a ROM (read-only memory), which would hold a look-up table for rounded results. For example, a ROM with l address lines would have as inputs the $(l - 1)$ least significant bits (out of the m bits) of the significand, and only the most significant bit out of the d extra bits, as depicted in Figure 4.6.

Number	$Round(x)$	Error	Number	$Round(x)$	Error
$X0.00$	$X0.$	0	$X1.00$	$X1.$	0
$X0.01$	$X0.$	$-1/4$	$X1.01$	$X1.$	$-1/4$
$X0.10$	$X0.$	$-1/2$	$X1.10$	$X1. + 1$	$+1/2$
$X0.11$	$X1.$	$+1/4$	$X1.11$	$X1. + 1$	$+1/4$

TABLE 4.7   The rounding errors for the round-to-nearest-even scheme with $d = 2$.

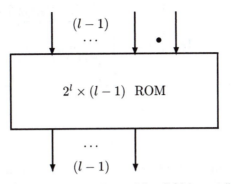

FIGURE 4.6   An implementation of the ROM rounding scheme.

The ROM in this figure has 2^l rows of $(l-1)$ bit each, containing a properly rounded $(l-1)$ result except when all $(l-1)$ low-order bits of the significand are 1's. In this case, the ROM returns all 1's (thus effectively truncating the result instead of rounding it) and avoids the full addition. If, for example, $l = 8$, the table look-up would be very fast and yet 255 out of 256 cases would be properly rounded. This rounding, for $l = 3$, is illustrated in Figure 4.7 [11].

The bias of ROM rounding for $l = 3$ and $d = 1$ can be calculated from Table 4.8. The sum of errors is $+1/2$ for the first three groups of size $2^d = 2$ and is $-1/2$ for the fourth group. The average bias is therefore $1/8$. In the general case, for given values of d and l, the average bias is $\frac{1}{2}[(\frac{1}{2})^d - (\frac{1}{2})^{l-1}]$. This bias converges to $\frac{1}{2}(\frac{1}{2})^d$ (e.g., $\frac{1}{4}$ for $d = 1$) when l is large enough, since ROM rounding converges to the round-to-nearest scheme. If the round-to-nearest-even modification is adopted, the bias of the modified ROM rounding converges to zero.

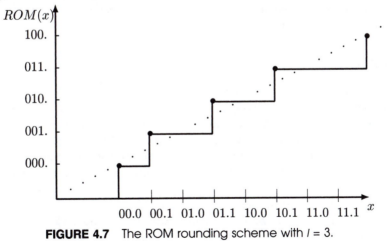

FIGURE 4.7   The ROM rounding scheme with $l = 3$.

Number	$ROM(x)$	Error	Number	$ROM(x)$	Error
$X00.0$	$X00.$	0	$X10.0$	$X10.$	0
$X00.1$	$X01.$	$+1/2$	$X10.1$	$X11.$	$+1/2$
$X01.0$	$X01.$	0	$X11.0$	$X11.$	0
$X01.1$	$X10.$	$+1/2$	$X11.1$	$X11.$	$-1/2$

TABLE 4.8   The rounding errors for the ROM rounding scheme with $l = 3$ and $d = 1$.

The IEEE floating-point standard [9] includes four rounding modes. The default is the round-to-nearest-even mode. The remaining three are directed roundings: round toward zero (truncate), round toward ∞ and round toward $-\infty$. The round toward $\pm\infty$ modes are useful for Interval Arithmetic in which each real number a is represented by two floating-point numbers a_1 and a_2 providing lower and upper bounds, respectively, for the real value. Thus, an operand is represented by an interval and all arithmetic operations operate on intervals [1]. For example, the add, subtract and multiply operations in interval arithmetic are defined as follows

$$[a_1, a_2] + [b_1, b_2] = [a_1 + b_1, a_2 + b_2]$$

$$[a_1, a_2] - [b_1, b_2] = [a_1 - b_2, a_2 - b_1]$$

$$[a_1, a_2] \times [b_1, b_2] = [\min\{a_1 b_1, a_1 b_2, a_2 b_1, a_2 b_2\}, \max\{a_1 b_1, a_1 b_2, a_2 b_1, a_2 b_2\}]$$

In these computations, the lower bound (e.g., $a_1 + b_1$ in addition) is rounded toward $-\infty$ while the upper bound ($a_2 + b_2$ in addition) is rounded toward ∞. The intervals calculated for the final results will provide an estimate on the accuracy of the computation.

4.6 GUARD DIGITS

When multiplying two significands, we obtain a double-length result, and clearly not all the extra digits are needed for proper rounding. A similar situation arises when adding or subtracting two numbers that do not have the same exponent. The question now is, What is the smallest number of extra digits that we need within the arithmetic unit? These extra digits are used for rounding and for postnormalization in the case when leading zeros are obtained. In what follows, we will first consider multiply/divide operations, and then add/subtract operations, since the former are simpler to handle.

As we have seen in Section 4.2, if signed-magnitude fractions are used as significands, then no extra digits for postnormalization are needed for division. A shift right operation may, however, be required. When multiplying two normalized fractional significands, at most one shift left is needed if $\beta = 2$ (k positions shift when $\beta = 2^k$). Therefore, one guard digit (in the radix β number system) is sufficient for postnormalization. A second guard digit is needed for rounding if the round-to-nearest scheme is adopted. Thus, a total of two guard digits suffice. These two digits are called the G (guard) and the R (round) digits. The same conclusion is reached when the significand is a signed-magnitude number in the range [1,2), as is the case in the IEEE standard. The proof of this statement is left as a exercise for the reader.

Implementing the round-to-nearest-even scheme requires an indicator to point out whether all the additional digits that were generated in the multiply operation are zero, in order to detect a tie. This indicator can be implemented as a single bit, which is the logical OR of all additional bits, and is known as the *sticky bit*. Thus, three bits, namely, G, R, and S (sticky) are sufficient, even for the round-to-nearest-even scheme. Computing the sticky bit when multiplying two significands does not require the generation of all the least significant bits of the product. The number of trailing zeros in the product of two binary significands is equal to the number of trailing zeros in the multiplier plus the number of trailing zeros in the multiplicand. Thus, the sticky bit should be set to 1 only if the expected number of trailing zeros in the double-length product is smaller than the number of the least significant product bits that are discarded. Other techniques for computing the sticky bit for the product are presented in [15].

The case of addition/subtraction of floating-point numbers is more complicated, especially when the final operation called for (after examining the sign bits) is subtraction. As before, we assume that the significands of the operands are normalized signed-magnitude fractions. Revisiting the subtract example in Section 4.2, it seems that all shifted-out digits of the subtrahend must be kept, and must participate in the subtract operation, in order to have all the necessary digits for the postnormalization step. This will require as many guard digits as the number of digits in the significand field, doubling the size of the significand adder/subtractor. However, if the significand of the subtrahend is shifted by more than one position to the right in the pre-alignment step, the resulting difference will have at most one leading zero. This implies that at most one of the shifted-out digits may be required for the postnormalization step.

Example 4.5

In the following subtract operation the significands of the two operands A and B are 12 bits long and the base is 2. Assume that the difference between the exponents is $E_A - E_B = 2$, requiring a 2-bit position shift of the subtrahend B in the pre-alignment step.

A	0.100000101100	00
B aligned	0.001100000001	10
$A - B$	0.010100101010	10
Postnormalization	0.101001010101	

Exactly the same result is obtained even if only one guard bit participates in the subtraction and generates the necessary borrow.   □

The situation is different if the most significant shifted-out bit is equal to zero, as illustrated in the following example.

Example 4.6

Consider the same two significands as in the previous example, but now assume that the difference between the exponents is $E_A - E_B = 6$. Thus, B's significand is shifted by six positions to the right to align it with A's significand.

A	0.100000101100	000000
B aligned	0.000000110000	000110
$A - B$	0.011111111011	111010
Postnormalization	0.111111110111	

If only one guard bit participated in the subtraction, then the four least significant bits of the result after the postnormalization step would be 1000 instead of 0111.   □

In the above subtraction, a long sequence of borrows is obtained, and it seems that we may need all the additional digits in B to guarantee that a borrow is generated. We might conclude that, in the worst case, we must double the number of digits. However, a careful analysis leads to the conclusion that it suffices to distinguish between two cases: (1) All additional bits (not including the guard bit) are zero, and (2) at least one of the additional bits is 1. To prove this, notice that all the extra digits in A are zeros (A was not preshifted); hence, the resulting three least significant bits in $A - B$ (011 in the above example) are independent of the exact position of the 1's in the extra digits of B. The only thing we need to know is whether a 1 was shifted out or not, and the sticky bit can be used for this purpose. If a 1 is shifted into it during the alignment, it will be set to one; otherwise, it remains zero. It is therefore again the logical OR of all the extra bits of B. The sticky bit participates in the subtraction and generates the necessary borrow. Thus, using the two digits, G and S, the previous subtraction looks like

		G	S
A	0.100000101100	0	0
B aligned	0.000000110000	0	1
$A - B$	0.011111111011	1	1
Postnormalization	0.111111110111		

The two bits, G and S, are sufficient for postnormalization. If, however, the round-to-nearest scheme is followed, an additional accurate bit is required. The sticky bit, which is only an indicator, cannot serve this purpose. Thus, three bits, namely, G, R, and S are required, as illustrated in the next example.

Example 4.7

Consider the following two significands, which are almost the same as those in the previous example, except for one bit in B, which is indicated in bold face. As before, assume that the difference between the exponents is $E_A - E_B = 6$. The correct result should be

A	0.100000101100	000000
B aligned	0.000000110000	010110
$A - B$	0.011111111011	101010
Postnormalization	0.111111110111	0

However, if we use only the G and S bits, we obtain

		G	S
A	0.100000101100	0	0
B aligned	0.000000110000	0	1
$A - B$	0.011111111011	1	1
Postnormalization	0.111111110111	1	

The round bit after the postnormalization step should be zero, and the sticky bit cannot be used for rounding. We must use the three bits, G, R, and S, as shown below:

		G	R	S
A	0.100000101100	0	0	0
B aligned	0.000000110000	0	1	1
$A - B$	0.011111111011	1	0	1
Postnormalization	0.111111110111	0		

The correct R bit, with a value of 0, is now available, and can be used in the round-to-nearest scheme. Note that if the round-to-nearest-even method is followed, the sticky bit, which is needed to detect a tie, is already available. This bit therefore, serves two purposes. □

If no postnormalization is required, then, for rounding purposes, we do not need the three bits G, R, and S, but should use only two bits, a round bit and a sticky bit. The original G bit can serve as an R bit, and the original R and S bits must be ORed in order to generate a new sticky bit. We are then left with two bits, which we call R and S, for the round-to-nearest-even procedure. This situation is illustrated in the following example.

Example 4.8

In the following subtraction, no postnormalization is needed. The difference between the exponents is $E_A - E_B = 6$.

		G	R	S
A	0.100001010100			
B	0.110000010001			
A	0.100001010100	0	0	0
B aligned	0.000000110000	0	1	1
$A - B$	0.100000100011	1	0	1
		R	S	
Before rounding	0.100000100011	1	1	
After round-to-nearest	0.100000100100			

$\square$

In the procedure above, if the new R bit is zero then no rounding is required and the sticky bit only indicates whether the final result is exact (if $S = 0$) or inexact (if $S = 1$). If $R = 1$ then the operation to be performed in the round-to-nearest-even procedure depends on the sticky bit and on the least-significant bit (denoted by L) of the resultant significand. If the sticky bit equals 1 then rounding must be performed by adding a *ulp* to the significand. If the sticky bit equals 0 then this is the tie case and only if $L = 1$ is rounding necessary. The above rules can be stated more succinctly by saying that the round-to-nearest-even scheme requires the addition of a *ulp* to the significand if the Boolean expression

$$R \cdot S + R \cdot \overline{S} \cdot L \;=\; R \cdot (S + L)$$

equals 1.

The addition of a *ulp* to the significand is sometimes needed even when a directed rounding mode is followed. For example, in the round toward $+\infty$ mode, a *ulp* must be added to the significand if the result is positive and either R or S equals 1. A similar situation occurs in the round toward $-\infty$ mode when the result is negative and the Boolean expression $R + S$ equals 1. The rules for all four rounding modes mandated by the IEEE floating-point standard are shown in Table 4.9.

LSB	R	S	Operation	$\overline{Error}$
0	0	0	$+0$	0
0	0	1	$+0$	$-0.25\ ulp$
0	1	0	$+0$	$-0.50\ ulp$
0	1	1	$+0.5\ ulp$	$+0.25\ ulp$
1	0	0	$+0$	0
1	0	1	$+0$	$-0.25\ ulp$
1	1	0	$+0.5\ ulp$	$+0.50\ ulp$
1	1	1	$+0.5\ ulp$	$+0.25\ ulp$
			Total	0

(a) Round-to-nearest-even scheme

R	S	Operation	$\overline{Error}$
0	0	$+0$	0
0	1	$+0$	$-0.25\ ulp$
1	0	$+0$	$-0.50\ ulp$
1	1	$+0$	$-0.75\ ulp$
		Total	$-0.375\ ulp$

(b) Round-to-zero scheme

Sign	R	S	Operation
$+$	0	0	$+0$
$+$	0	1	$+1\ ulp$
$+$	1	0	$+1\ ulp$
$+$	1	1	$+1\ ulp$
$-$	0	0	$+0$
$-$	0	1	$+0$
$-$	1	0	$+0$
$-$	1	1	$+0$

(c) Round-to-plus-infinity scheme

Sign	R	S	Operation
$-$	0	0	$+0$
$-$	0	1	$+1\ ulp$
$-$	1	0	$+1\ ulp$
$-$	1	1	$+1\ ulp$
$+$	0	0	$+0$
$+$	0	1	$+0$
$+$	1	0	$+0$
$+$	1	1	$+0$

(d) Round-to-minus-infinity scheme

TABLE 4.9   The rules of all four rounding schemes.

The addition of a *ulp* that comes after the two significands have been added substantially increases the execution time of a floating-point add/subtract operation. The extra delay due to rounding can be avoided because all three guard bits are known before the significands are being added. Thus, the addition of 1 to L can be done at the same time that the significands are added. The exact position of the L bit is not known yet, since a postnormalization may be required. However, the L bit has only two possible positions and two adders can therefore be used in parallel, providing the correctly rounded results for both cases [7]. This can also be achieved using one adder as described in the next section.

4.7 FLOATING-POINT ADDERS

The procedure followed when adding two floating-point numbers, depicted in Figure 4.1, includes a large number of steps which are executed sequentially.

However, a careful examination of the procedure reveals that not all of these steps must be executed for each operation, and that some of the steps can be executed in parallel.

To this end, we distinguish between effective addition and effective subtraction [2]. The effective operation depends on the sign bits of the two operands and the instruction to be executed. For effective addition, we first calculate the exponent difference to determine the alignment shift. We then shift the significand of the smaller operand, and add these aligned significands. The result of the addition can overflow by at most one bit position. Thus, a postnormalization shift, which would be time-consuming, is not needed. The single bit overflow can be easily detected and, if it is found, a 1-bit normalization of the result is performed using a multiplexor.

To eliminate the need for an increment operation in the rounding step, the above-mentioned significand adder is designed to produce two simultaneous results, *sum* and *sum+1*. An adder capable of producing the two results *sum* and *sum+1* is sometimes called a compound adder [13], and it can be implemented in various ways, including carry-look-ahead and conditional sum (see Chapter 5). For the IEEE round-to-nearest-even rounding mode, we use the rounding bits to determine which of these two results should be selected. Note that these two results (i.e., *sum* and *sum+1*) are sufficient even if a single bit overflow occurs. In case of an overflow, the 1 will be added in the R bit position (instead of in the least significant bit (LSB) position), and since $R = 1$ if rounding is needed, a carry will be propagated into the LSB position to generate the correct value of *sum+1*. However, for the two directed rounding modes (round to $\pm\infty$), the R bit is not necessarily 1 and, as a result, *sum+2* may be needed in the case of the 1-bit overflow.

In effective subtraction, massive cancellation of the most significant bits may occur, resulting in the need for a lengthy postnormalization step. However, this can happen only when the exponents of the two operands are close (i.e., the difference is less than or equal to 1), in which case the pre-alignment step can be eliminated. It therefore makes sense to implement effective subtraction as two separate procedures: one for the case where the exponents are close, and one for the case where the exponent difference is greater than 1. In the *CLOSE* case only a postnormalization shift may be required, while in the *FAR* case only a pre-alignment shift may be needed. The required steps in the *CLOSE* and *FAR* cases are shown in Table 4.10.

In the *CLOSE* case we first predict the exponent difference, based on the two least significant bits of the operands, to allow the subtraction of the significands to start as soon as possible. If the predicted difference is zero, a subtract will be executed with no alignment. If the predicted difference is ±1, the significand of the smaller operand is shifted once to the right (using a multiplexor) and then subtracted from the other significand. At the same time, the true exponent

Step	CLOSE	FAR
1	Predict exponent	Subtract exponents
2	Subtract significands Predict number of leading zeroes	Align significands
3	Postnormalization	Subtract significands
4	Select properly rounded result or negate result	Select properly rounded result

TABLE 4.10 The steps in the CLOSE and FAR cases.

difference is calculated. If this difference is greater than 1, the above procedure is aborted and the *FAR* procedure is followed. If, however, the true exponent difference is less than or equal to ± 1, the *CLOSE* procedure is continued. In parallel with the subtraction, the number of leading zero bits is predicted, in order to determine the number of shift positions in the postnormalization step. The normalization of the significand and the corresponding exponent adjustment are done next. Finally, rounding is performed in step 4. As mentioned above, rounding can be accomplished by precomputing *sum* and *sum+1* and then selecting the one which is properly rounded. In step 4 a negation of the result may also be needed. Since the subtract operation is almost always executed so that the smaller (pre-shifted) significand is subtracted from the larger one, the result of the subtraction is usually positive and negation is not required. Note that the sign of the final result is determined by the sign of the largest operand. Only in the case where the exponents of the two operands are equal, the result of the significand subtraction may be negative (represented in two's complement), requiring a negation step. However, in such a case, no pre-alignment is performed, and consequently, no guard bits are generated (the result is exact) and hence no rounding is necessary. Thus, the negation and rounding steps are mutually exclusive.

In the *FAR* case the exponent difference is calculated first, and then the significand of the smaller operand is shifted to the right to align it with the other significand. The shifted-out bits are used to set the sticky bit. The smaller significand is now subtracted from the larger significand, with the result being either normalized or requiring a single-bit-position left-shift, which is accomplished using a multiplexor. In step 4, rounding is performed.

We conclude this section with a brief description of the leading zeroes prediction circuit. This circuit should predict the position of the leading non-zero bit in the result of the subtract operation before the subtraction is completed. This would allow us to execute the postnormalization shift immediately following the subtraction. One way to achieve this is to examine the bits of the two operands (of the subtract operation) in a serial fashion, starting with the most significant bits to determine the position of the first 1 [8]. This serial operation can be accelerated using a parallel scheme similar to the carry-look-ahead technique (see Chapter 5).

Another way to predict the position of the leading 1 is to generate in parallel a set of intermediate bits e_i such that $e_i = 1$ if the corresponding bits a_i and b_i of the two operands are identical and the previous bits, i.e., a_{i-1} and b_{i-1}, allow the propagation of the expected carry (i.e., at least one of the bits a_{i-1} and b_{i-1} equals 1). A carry is expected since the subtract operation is executed by forming the one's complement of the subtrahend and forcing a carry into the least significant position. (In the notation used in Chapter 5, the Boolean expression for e_i is $e_i = \overline{a_i \oplus b_i}(a_{i-1}+b_{i-1})$ where b_i is the complement of the original subtrahend bit.) In other words, $e_i = 1$ if a carry is allowed to propagate to position i. The corresponding ith bit of the correct result will also be equal to 1 unless the forced carry from the least significant position did not propagate to position $i - 1$; in such a case, the correct result will have a 1 in position $i - 1$ instead. Thus, the position of the leading 1 in the result is either identical to the position of the leading 1 in the sequence of the e_i bits, or it is one position to the right. We may therefore count the number of leading zeroes in the sequence of the e_i bits and provide this count to the barrel shifter executing the postnormalization shift. After this, at most one bit position correction shift (to the left) will be required [18]. A comparison of several methods for predicting the position of the leading bit appears in [16].

4.8 EXCEPTIONS

We will concentrate in this section on the exceptions specified in the IEEE standard [9]. The 754 standard defines five types of exceptions: overflow, underflow, division-by-zero, invalid operation and inexact result. The first three exceptions are found in almost all floating-point systems. Only the last two are peculiar to the IEEE standard.

When an exception occurs, a status flag is set and a specified result is generated (e.g., a correctly signed ∞ when a division-by-zero occurs). The status flag should remain set until explicitly cleared. The IEEE standard recommends the implementation of a separate trap-enable bit for each exception. If this bit is on when the corresponding exception occurs then, in addition to setting the status flag, the user trap handler is called. Sufficient information must be provided by the floating-point unit to the trap handler to allow it to take the appropriate action, e.g., exact identification of the operation which caused the exception.

Overflow.   The overflow exception flag is set whenever the exponent of the result exceeds the largest value allowed in the result's format. For example, in the single-precision format an overflow occurs if $E > 254$. The result, when the corresponding trap is disabled, is determined by the sign of the intermediate (overflowed) result and the rounding mode as follows:

1. In the round-to-nearest-even mode an ∞ with the sign of the intermediate result is generated.

2. In the round toward 0 mode the largest representable number with the sign of the intermediate result is generated.

3. In the round toward $-\infty$ mode the largest representable number with a plus sign is generated if the intermediate result is positive. Otherwise the final result is set to $-\infty$.

4. In the round toward ∞ mode the largest representable number with a minus sign is generated if the intermediate result is negative. Otherwise the final result is set to $+\infty$.

If the overflow trap is enabled then the trap handler receives the intermediate result divided by 2^a and then rounded where a is 192 or 1536 for the single- and double-precision format, respectively. This scaling adjustment was chosen in order to translate the overflowed result as nearly as possible to the middle of the exponent range so that it can be used in subsequent operations with less risk of causing further exceptions. For example, when multiplying the number 2^{127} (for which $E = 254$ in the single-precision format) by 2^{127}, the overflowed product has an exponent of $E = 254 + 254 - 127 = 381$ after being adjusted by 127, as is normally done for the multiply operation. This result clearly overflows since $E > 254$. If this product is then scaled (multiplied) by 2^{-192}, the resulting exponent becomes $E = 381 - 192 = 189$ which represents the "true" value of $189 - 127 = 62$. This scaled intermediate value has a smaller risk of causing further exceptions. Consider now the case where relatively "small" operands result in an overflow. If we multiply the number 2^{64} (for which $E = 191$ in the single-precision format) by 2^{65} ($E = 192$), the overflowed product has an exponent of $E = 191 + 192 - 127 = 256$. If this exponent is then adjusted by 192, we obtain $E = 256 - 192 = 64$ which represents the "true" value of $64 - 127 = -63$. Conversions (e.g., from decimal to binary) are handled differently, see [9].

Underflow.   The conditions under which the underflow flag is set depend on whether the underflow trap is enabled or disabled. If the underflow trap is enabled the underflow exception flag is set whenever the result is a nonzero number between $-2^{E_{min}}$ and $2^{E_{min}}$. Recall that E_{min} is -126 for the single-precision format and -1022 for the double-precision format. The intermediate result delivered to the underflow trap handler is the infinitely precise result multiplied by 2^a and then rounded. As for overflow, a equals 192 or 1536 for the single- or double-precision format, respectively. Conversions are also handled differently [9].

If however, the underflow trap is disabled, denormalized numbers are allowed. The underflow exception flag is then set only when an extraordinary loss of accuracy occurs while representing the intermediate result (which has a nonzero value between $\pm 2^{E_{min}}$) as a denormalized number. Such a loss of accuracy occurs when either the guard bit or the sticky bit is nonzero. These indicate an inexact result. In an arithmetic unit where denormalized numbers are not implemented the delivered result is either zero or $\pm 2^{E_{min}}$.

Example 4.9

Suppose the underflow exception trap is disabled and denormalized numbers have been implemented. If we multiply 2^{-65} by 2^{-65} the resulting exponent is $E = (127-65)+(127-65)-127 = -3$. Since $E < 1$ we cannot represent the product as a normalized number. Instead, we represent the result 2^{-130} as the denormalized number $0.0001 \cdot 2^{-126}$, i.e., $f = .0001$ and $E = 0$. No underflow exception flag is set.

If the second operand is $(1 + ulp)2^{-65}$ (rather than 2^{-65}), the correct product is $(1+ulp)2^{-130}$ which, when converted to a denormalized number, yields $f = .0001$ and $E = 0$ as before but now the sticky bit is equal to 1. Therefore, this is an inexact result and the underflow exception flag is set.

□

Division by zero.   The division-by-zero exception flag is set whenever the divisor is zero and the dividend is a finite nonzero number. When the corresponding trap is disabled the result must be a correctly signed ∞.

Invalid operation.   The invalid operation flag is set if an operand is invalid for the operation to be performed. The result, when the invalid operation trap is disabled, is a quiet NaN. Examples of invalid operations are [9]:

1. Multiplying 0 by ∞.

2. Dividing 0 by 0 or ∞ by ∞.

3. Adding $+\infty$ and $-\infty$.

4. Finding the square root of a negative operand.

5. Calculating the remainder x REM y where y is zero or x is infinite.

6. Any operation on a signaling NaN.

Inexact result.   The inexact result flag is set if the rounded result is not exact or if it overflows without an overflow trap. A rounded result is exact only when both the guard bit and the sticky bit are equal to zero. This implies that no precision was lost when performing the rounding. The purpose of the inexact result flag is to allow integer calculations to be performed in a floating-point unit.

4.9 ROUND-OFF ERRORS AND THEIR ACCUMULATION

The need to perform rounding in floating-point operations, even with the best rounding scheme, results in errors that tend to accumulate as the number of operations increases. The relative round-off error in a floating-point operation is denoted by ϵ and is defined by the equation

$$\epsilon = \frac{Fl(x \star y) - (x \star y)}{(x \star y)} \tag{4.8}$$

where $\star$ is any one of the floating-point arithmetic operations $+, -, \times,$ or $/$, and $Fl(x \star y)$ is the correctly rounded or truncated result of the operation $(x \star y)$. A more convenient form of Equation (4.8) is

$$Fl(x \star y) = (x \star y) \cdot (1 + \epsilon). \tag{4.9}$$

Upper bounds for the relative error ϵ can be derived for the different round-off schemes. For truncation, the absolute error can be almost as large as the least-significant digit of the significand. The worst case for the relative error is when the normalized result assumes its smallest possible value. Therefore,

$$|\epsilon_{trunc}| \leq \frac{2^{-m}}{1/2} = 2^{-m+1}.$$

For the round-to-nearest scheme, the maximum absolute error is only half of ulp, and consequently

$$|\epsilon_{round}| \leq \frac{1}{2} 2^{-m+1} = 2^{-m}.$$

The above formulas provide only upper bounds for the relative error. What might be more important is the exact distribution of the relative error within its bounds. This distribution has been studied (e.g., in [20]) and the following density function of ϵ_{trunc} (see Figure 4.8(a)) has been derived:

$$f_{\epsilon_{trunc}}(\epsilon_0) = \begin{cases} 2^{m-1}/\ln 2 & \text{if} \quad 0 \leq \epsilon_0 < 2^{-m} \\ (\frac{1}{\epsilon_0} - 2^{m-1})/\ln 2 & \text{if} \quad 2^{-m} \leq \epsilon_0 < 2^{-m+1} \end{cases} \tag{4.10}$$

In other words, the relative errors for truncation are uniformly distributed in the region $[0, 2^{-m}]$ and reciprocally distributed in the region $[2^{-m}, 2^{-m+1}]$. The

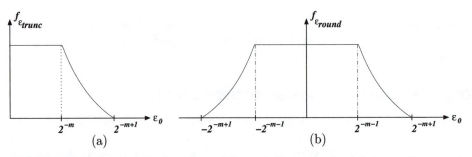

FIGURE 4.8   The density function of the relative error for (a) truncation, and (b) rounding.

average value of the relative error can now be calculated. This calculation results in

$$\overline{\epsilon_{trunc}} = 2^{-m-1}/\ln 2 \approx 0.72 \cdot 2^{-m}.$$

The density function of ϵ_{round} (see Figure 4.8(b)) is

$$f_{\epsilon_{round}}(\epsilon_0) \;=\; \begin{cases} 2^{m-1}/\ln 2 & \text{if} \quad -2^{-m-1} \le \epsilon_0 \le 2^{-m-1} \\ \left(\frac{1}{2|\epsilon_0|} - 2^{m-1}\right)/\ln 2 & \text{if} \quad 2^{-m-1} < |\epsilon_0| < 2^{-m} \end{cases} \qquad (4.11)$$

The relative errors, when rounding, are uniformly distributed in the region $[-2^{-m-1}, 2^{-m-1}]$ and reciprocally distributed elsewhere. Unlike Equation (4.10), the density function in Equation (4.11) is symmetric with respect to $\epsilon_0 = 0$ and, as a result, the average relative error is 0. The analytically derived expressions in Equations (4.10) and (4.11) were shown to be in very good agreement with empirical results.

The above analysis concentrated on the round-off errors occurring in a single floating-point operation. It did not indicate how these errors could accumulate in subsequent operations. Consider, for example, two intermediate results A_1 and A_2, which are to be added. Denote by ϵ_1 and ϵ_2 their corresponding relative errors, satisfying

$$A_1 = A_1^c(1 + \epsilon_1), \qquad A_2 = A_2^c(1 + \epsilon_2)$$

where A_1^c and A_2^c denote the "correct" values of A_1 and A_2, respectively. If we focus on accumulated error and assume that no new error will be introduced in the addition, then the relative error of the sum $A_1 + A_2$ becomes

$$\frac{A_1^c \cdot \epsilon_1 + A_2^c \cdot \epsilon_2}{A_1^c + A_2^c} = \frac{A_1^c}{A_1^c + A_2^c} \cdot \epsilon_1 + \frac{A_2^c}{A_1^c + A_2^c} \cdot \epsilon_2$$

Thus, the relative error of the sum is a weighted average of the relative errors of the operands. If the two operands are positive, then the above error will be dominated by the error in the larger operand.

A more severe case of error accumulation may occur in the subtract operation $A_1 - A_2$. The relative error in this case, under the same assumption as above, equals

$$\frac{A_1^c}{A_1^c - A_2^c} \cdot \epsilon_1 - \frac{A_2^c}{A_1^c - A_2^c} \cdot \epsilon_2$$

If the operands A_1 and A_2 are positive numbers close to each other, the accumulated relative error could increase significantly, especially if the two relative errors ϵ_1 and ϵ_2 have opposite signs, resulting in a very substantial inaccuracy.

The accumulation of errors when a sequence of several floating-point operations are performed depends on the particular set of operations performed; in other words, it depends on the specific application. Therefore, the accumulation of errors in the general case cannot be analyzed and simulative studies must be used instead (e.g., [10], [12]). As might be expected, these studies have shown that in most cases, the accumulated relative error when truncation is used is higher than that when the round-to-nearest scheme is employed.

4.10 EXERCISES

4.1. Consider the following 40-bit floating-point format: A sign bit, a 29-bit normalized (fractional) significand in two's complement form, a 10-bit excess 512 exponent, and base 2 for the exponent. Determine the bit patterns of the smallest and largest positive and negative numbers, and find their value. Also, find the distance between any two consecutive numbers. How many different values can you represent using this format?

4.2. Find the normalized internal machine representations of the following floating-point numbers in the short IBM format, in the short DEC format and in the short IEEE format: (a) $+0.25 \cdot 2^{+31}$ (b) $-31.75 \cdot 2^{-7}$

4.3. (a) Show that there are 1.875 times as many normalized floating-point numbers with an exponent base of 16 as those with an exponent base of 2. Both use the same number of bits in the exponent and significand fields.
(b) What is the ratio of representable normalized numbers for $\beta = 4$?

4.4. How many denormalized numbers are there in the short (32-bit) IEEE format, and what is their range? Compare it to the number of normalized values with the fixed exponent 1 (including bias).

4.5. Which of the following properties are satisfied in systems with denormalized numbers, and which are satisfied even in systems with no denormalized numbers:
(a) $x \neq y$ implies $x - y \neq 0$. (b) $(x - y) + y \simeq x$ (with a rounding error).
(c) For a normalized number x, if $1/x \neq 0$ then $\frac{1}{1/x} \simeq x$ (with a rounding error).

4.6. Prove that the average bias of the ROM rounding scheme is $\frac{1}{2}[(\frac{1}{2})^d - (\frac{1}{2})^{l-1}]$.

4.7. Four rounding schemes are supported by the IEEE standard: Round-to-nearest-even, round toward zero (truncate), round toward infinity, and round toward $-\infty$. Calculate the bias of all four rounding schemes.

4.8. For the four rounding schemes supported by the IEEE standard (see problem 7) show the final rounded results in the following three cases:

s	$exponent$	$fraction$	$guard$
0	00011111	11111111111111111111111	1
0	11111110	11111111111111111111111	1
1	11111110	11111111111111111111111	1

4.9. Based on the results of problem 8, what is the advantage of the round-to-nearest-even scheme? What is the disadvantage of the round-to-nearest-even scheme (when implementation is considered), which can be avoided if a round-to-nearest-odd scheme is adopted?

4.10. Write down the postnormalization steps that might be needed when performing addition, subtraction, multiplication, and division with two floating-point operands in the IEEE short format. Indicate how many guard digits are needed in each case.

4.11. Show the result of the following operations on numbers in the IEEE short format in all four rounding schemes (see problem 7). The operands are given in the hexadecimal notation.
(**a**) 3F80 0000 + 0080 0000   (**b**) 3F80 0000 − 3F7F FFFF
(**c**) 3F80 0000 + 3380 0000   (**d**) 3F80 0000 − 0080 0000
(**e**) 4000 0001 × 4000 0001   (**f**) 4000 0000 − 3380 0000

4.12. Two normalized floating-point numbers A and B in the short IEEE format were added, and the result was equal to A. Does this imply that $B = 0$?

4.13. Given a floating-point number A with an exponent E_A (in any format), its successor has either the same exponent or the exponent $E_A + 1$. Is the distance between A and its successor the same in both cases?

4.14. (**a**) Compare the error involved in the serial evaluation of the product of four numbers, performed as $(((A_1 \times A_2) \times A_3) \times A_4)$ to that of its parallel evaluation performed as $((A_1 \times A_2) \times (A_3 \times A_4))$. Decide whether one of these methods has a smaller upper bound for the error when forming the product of n numbers.
(**b**) Repeat (a) for the sum of four numbers, then n numbers. Can we get lower error bounds if we know that the numbers are in some order; e.g., ascending order?

4.15. Prove that the optimal way to implement a two-level combinatorial shifter for k bits, where $k = m^2$, is for the first level to shift by multiples of m, and the second level to shift from 0 to m. Assume that the delay is proportional to the number of destinations for each line in the two levels. Can you generalize this result for any value of k?

4.16. Another way to implement a radix-4 combinatorial shifter for 53 bits is by restricting the number of destinations for every bit in each level to 4 but allow more than two levels. How many levels will such a combinatorial shifter have? How many levels will a radix-r combinatorial shifter for m bits have if the number of destinations for every bit in each level is restricted to $r+1$?

4.11 REFERENCES

[1]   G. ALEFELD and J. HERZBERGER, *Introduction to interval computations,* Academic Press, NY, 1983.

[2]   B. J. BENSCHNEIDER, *et al.* "A pipelined 50-MHz CMOS 64-bit floating-point arithmetic processor," *IEEE Journal on Solid-State Circuits, 24* (October 1989), 1317-1323.

[3]   R. P. BRENT, "On the precision attainable with various floating-point number systems," *IEEE Trans. on Computers, C-22* (June 1973), 601-607.

[4]   W. J. CODY, JR., "Static and dynamic numerical characteristics of floating-point arithmetic," *IEEE Trans. on Computers, C-22* (June 1973), 598-601.

[5]   W. J. CODY, JR., "Analysis of proposals for the floating-point standard," *Computer,* (March 1981), 63-69.

[6]   G. EVEN and P.-M. SEIDEL, "A comparison of three rounding algorithms for IEEE floating-point multiplication," *IEEE Trans. on Computers, 49* (July 2000), 638-650.

[7]   D. GOLDBERG, "Computer arithmetic," in *Computer architecture: A quantitative approach,* D. A. Patterson and J. L. Hennessy, Morgan Kaufmann, CA, 1996.

[8]   E. HOKENEK, R. MONTOYE and P. COOK, "Second-generation RISC floating-point with multiply-add fused," *IEEE Journal of Solid-State Circuits, 25* (October 1990), 1207-1213.

[9]   "IEEE standard for binary floating-point arithmetic," ANSI/IEEE 754-1985, also in *Computer,* 14 (March 1981), 51-62.

[10] D. J. KUCK, *et al.* "Analysis of rounding methods in floating-point arithmetic," *IEEE Trans. on Computers, C-26* (7) (July 1977) pp. 643-650.

[11] D. J. KUCK, *The Structure of computers and computations,* vol. 1, Wiley, New York, 1978, chap. 3.

[12] J. D. MARASA and D. W. MATULA, "A simulative study of correlated error propagation in various finite-precision arithmetics," *IEEE Trans. on Computers, C-22* (June 1973), 587-597.

[13] S. F. OBERMAN, H. AL-TWAIJRY, and M. J. FLYNN, "The SNAP project: Design of floating-point arithmetic units," *Proc. of 13th Symp. on Computer Arithmetic* (July 1997), 156-165.

[14] V. PENG, S. SAMUDRALA and M. GAVRIELOV, "On the implementation of shifters, multipliers and dividers in VLSI floating-point units," *Proc. of 8th Symp. on Computer Arithmetic* (May 1987), 95-102.

[15] M. R. SANTORO, G. BEWICK and M. A. HOROWITZ, "Rounding algorithms for IEEE multipliers," *Proc. 9th Symp. on Computer Arithmetic,* 1989, 176-183.

[16] M. S. SCHMOOKLER and K. J. NOWKA, "Leading zero anticipation and detection - A comparison of methods," *Proc. 15th Symp. on Comp. Arithmetic,* 2001, 7-12.

[17] P. H. STERBENZ, *Floating-point computation,* Prentice Hall, Englewood Cliffs, NJ, 1974.

[18] H. SUZUKI, H. MORINAKA, *et al.*, "Leading-zero anticipatory logic for high-speed floating point addition," *IEEE Journal on Solid-State Circuits, 31* (August 1996), 1157-1164.

[19] D. W. SWEENEY, "An analysis of floating-point addition," *IBM Systems Journal, 4* (1965) 31-42.

[20] N. TSAO, "On the distribution of significant digits and roundoff errors," *Communications of the ACM, 17* (May 1974), 269-271.

[21] J. M. YOHE, "Rounding in floating-point arithmetic," *IEEE Trans. on Computers, C-22* (June 1973) 577-586.

5

FAST ADDITION

5.1 RIPPLE-CARRY ADDERS

The addition of two operands is the most frequent operation in almost any arithmetic unit. A two-operand adder is used not only when performing additions and subtractions, but also often employed when executing more complex operations like multiplication and division. Consequently, a fast two-operand adder is essential.

The most straightforward implementation of a parallel adder for two operands $x_{n-1}, x_{n-2} \cdots, x_0$ and $y_{n-1}, y_{n-2} \cdots, y_0$ is through the use of n basic units called *full adders*. A full adder (FA) is a logical circuit that accepts two operand bits, say x_i and y_i, and an incoming carry bit, denoted by c_i, and then produces the corresponding sum bit, denoted by s_i, and an outgoing carry bit, denoted by c_{i+1}. As this notation suggests, the outgoing carry c_{i+1} is also the incoming carry for the subsequent FA, which has x_{i+1} and y_{i+1} as input bits. The FA is a combinational digital circuit implementing the binary addition of three bits through the following Boolean equations:

$$s_i = x_i \oplus y_i \oplus c_i \tag{5.1}$$

where $\oplus$ is the exclusive-or operation, and

$$c_{i+1} = x_i \cdot y_i + c_i \cdot (x_i + y_i) \tag{5.2}$$

where $x_i \cdot y_i$ is the AND operation, $x_i \wedge y_i$, and $x_i + y_i$ is the OR operation, $x_i \vee y_i$.

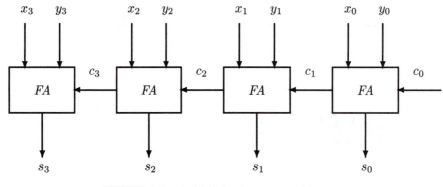

FIGURE 5.1   A 4-bit ripple-carry adder.

A parallel adder consisting of FAs for $n = 4$ is depicted in Figure 5.1. In a parallel arithmetic unit, all $2n$ input bits (x_i and y_i) are usually available to the adder at the same time. However, the carries have to propagate from the FA in position 0 (the position of the FA whose inputs are x_0 and y_0) to position i in order for the FA in that position to produce the correct sum and carry-out bits. In other words, we need to wait until the carries *ripple* through all n FAs before we can claim that the sum outputs are correct and may be used in further calculations. Because of this, the parallel adder shown in Figure 5.1 is called a *ripple-carry adder.* Note that the FA in position i, being a combinatorial circuit, will see an incoming carry $c_i = 0$ at the beginning of the operation, and will accordingly produce a sum bit s_i. The incoming carry c_i may change later on, resulting in a corresponding change in s_i. Thus, a ripple effect can be observed at the sum outputs of the adder as well, continuing until the carry propagation is complete. Also, notice that in an add operation, the incoming carry in position 0, c_0, is always zero and as a result the FA in this position can be replaced by a simpler unit capable of adding only two bits. Such a circuit exists and is called a half adder (HA), and its Boolean equations can be obtained from Equations (5.1) and (5.2) by setting c_i equal to 0. Still, an FA is frequently used to enable us to add a 1 in the least-significant position (*ulp*). This is needed to implement a subtract operation in the two's complement method. Here, the subtrahend is complemented and then added to the minuend. This is accomplished by taking the one's complement of the subtrahend and adding a *forced* carry to the FA in position 0 by setting $c_0 = 1$.

Example 5.1

Consider the following two operands for the adder in Figure 5.1: $x_3, x_2, x_1,$ $x_0 = 1111$ and $y_3, y_2, y_1, y_0 = 0001$. Δ_{FA} denotes the operation time (delay) of an FA, assuming that the delays associated with generating the sum output and the carry-out are equal. This may be the case if, for example, both circuits use a two-level gate implementation. The following diagram shows the sum and carry signals as a function of the time, T, measured in Δ_{FA} units:

$T = 0$		1111
	$+$	0001
$T = \Delta_{FA}$	Carry	0001
	Sum	1110
$T = 2\Delta_{FA}$	Carry	0011
	Sum	1100
$T = 3\Delta_{FA}$	Carry	0111
	Sum	1000
$T = 4\Delta_{FA}$	Carry	1111
	Sum	0000

This is the longest carry propagation chain that can occur when adding two 4-bit numbers. In synchronous arithmetic units, the time allowed for the adder's operation must be the worst-case delay, which is, in the general case, $n \cdot \Delta_{FA}$. This means that the adder is assumed to produce the correct sum after this fixed delay regardless of the actual carry propagation time, which might be very short, as in 0101+0010.

Consider now the subtract operation $0101 - 0010$, which is performed by adding the two's complement of the subtrahend to the minuend. The two's complement is formed by taking the one's complement of 0010, 1101, and setting the forced carry c_0 to 1, yielding 0011.   □

It is clear that the long carry propagation chains must be dealt with in order to speed up the addition. Two main approaches can be envisioned: One is to reduce the carry propagation time; the other is to detect the completion of the carry propagation and avoid wasting time while waiting for the fixed delay (of $n \cdot \Delta_{FA}$ for ripple-carry adders) unless absolutely necessary. Clearly, the second approach leads to a variable addition time, which may be inconvenient in a synchronous design. We will therefore concentrate on the first approach and study several schemes for accelerating carry propagation. The technique for detection of carry completion is left as an exercise to the reader.

5.2 CARRY-LOOK-AHEAD ADDERS

The most commonly used scheme for accelerating carry propagation is the *carry-look-ahead* scheme. The main idea behind carry-look-ahead addition is an attempt to generate all incoming carries in parallel (for all the $n - 1$ high order FAs) and avoid the need to wait until the correct carry propagates from the stage (FA) of the adder where it has been generated. This can be accomplished in principle, since the carries generated and the way they propagate depend only on the digits of the original numbers $x_{n-1}x_{n-2}\cdots x_0$ and $y_{n-1}y_{n-2}\cdots y_0$. These digits are available simultaneously to all stages of the adder and, consequently, each stage can have all the information it needs in order to calculate the correct value of the incoming carry and compute the sum bit accordingly. This, however, would require an inordinately large number of inputs to each stage of the adder, rendering this approach impractical.

One may reduce the number of inputs at each stage by extracting the information needed from the input digits to determine whether new carries will be generated and whether they will be propagated. To this end, we will study in detail the generation and propagation of carries.

There are stages in the adder in which a carry-out is generated regardless of the incoming carry, and as a result, no additional information on previous input digits is required. These are the stages for which $x_i = y_i = 1$. There are other stages that are only capable of propagating the incoming carry; i.e., $x_iy_i = 10$, or $x_iy_i = 01$. Only a stage in which $x_i = y_i = 0$ cannot propagate a carry. To assimilate the information regarding the generation and propagation of carries, we define the following logic functions using the AND and OR operations. Let $G_i = x_i \cdot y_i$ denote the *generated carry* and let $P_i = x_i + y_i$ denote the *propagated carry*. As a result, the Boolean expression (5.2) for the carry-out can be rewritten as

$$c_{i+1} = x_iy_i + c_i(x_i + y_i) = G_i + c_i \cdot P_i.$$

Substituting $c_i = G_{i-1} + c_{i-1}P_{i-1}$ in the above expression yields

$$c_{i+1} = G_i + G_{i-1}P_i + c_{i-1}P_{i-1}P_i.$$

Further substitutions result in

$$
\begin{aligned}
c_{i+1} =\ & G_i + G_{i-1}P_i + G_{i-2}P_{i-1}P_i + c_{i-2}P_{i-2}P_{i-1}P_i \ =\ \cdots \\
=\ & G_i + G_{i-1}P_i + G_{i-2}P_{i-1}P_i + \cdots + c_0P_0P_1\cdots P_i.
\end{aligned}
\tag{5.3}
$$

This type of expression allows us to calculate all the carries in *parallel* from the original digits $x_{n-1}x_{n-2}\cdots x_0$ and $y_{n-1}y_{n-2}\cdots y_0$ and the forced carry c_0. For example, for a 4-bit adder, the carries are

$$
\begin{aligned}
c_1 &= G_0 + c_0P_0, \\
c_2 &= G_1 + G_0P_1 + c_0P_0P_1, \\
c_3 &= G_2 + G_1P_2 + G_0P_1P_2 + c_0P_0P_1P_2, \\
c_4 &= G_3 + G_2P_3 + G_1P_2P_3 + G_0P_1P_2P_3 + c_0P_0P_1P_2P_3.
\end{aligned}
\tag{5.4}
$$

If this is done for all stages of the adder, then for each stage a Δ_G delay is required to generate all P_i and G_i, where Δ_G is the delay of a single gate.

A delay of $2\Delta_G$ is then needed to generate all c_i (assuming a two-level gate implementation) and another $2\Delta_G$ to generate the sum digits, s_i, in parallel (again, assuming a two-level gate implementation). Hence, a total of $5\Delta_G$ time units is needed, regardless of n, the number of bits in each operand. However, for a large value of n, say, $n = 32$, an extremely large number of gates is needed and, more importantly, gates with a very large fan-in are required (fan-in is the number of gate inputs, and is equal to $n + 1$ in this case). Therefore, we must reduce the span of the look-ahead at the expense of speed. We may divide the n stages into groups and have a separate carry-look-ahead in each group. The groups can then be interconnected by the ripple-carry method. Dividing the adder into equal-sized groups has the additional benefit of modularity, requiring the detailed design of only a single integrated circuit. A group size of 4 has been commonly used, and ICs capable of adding two sequences, each consisting of four digits with carry-look-ahead, are available. Size 4 was selected because it is a common factor of most word sizes, and also because of technology-dependent constraints (e.g., the available number of input/output pins).

For n bits and groups of size 4, there are $n/4$ groups. To propagate a carry through a group once the P_i's, G_i's, and c_0 are available, we need $2\Delta_G$ time units. Thus, $1\Delta_G$ is needed to generate all P_i and G_i, $(n/4) \cdot 2\Delta_G$ are needed to propagate the carry through all bits, and an additional delay of $2\Delta_G$ is needed to generate the sum outputs, for a total of $(2\frac{n}{4} + 3)\Delta_G = (\frac{n}{2} + 3)\Delta_G$. This is almost a fourfold reduction in delay compared to the $2n\Delta_G$ delay of a ripple-carry adder.

We may further speed up the addition by providing a carry-look-ahead over groups in addition to the internal look-ahead within the group. We define a *group-generated carry*, G^*, and a *group-propagated carry*, P^*, for a group of size 4 as follows: $G^* = 1$ if a carry-out (of the group) is generated internally and $P^* = 1$ if a carry-in (to the group) is propagated internally to produce a carry-out (of the group). The Boolean equations for these carries are

$$
\begin{aligned}
G^* &= G_3 + G_2 P_3 + G_1 P_2 P_3 + G_0 P_1 P_2 P_3, \\
P^* &= P_0 P_1 P_2 P_3.
\end{aligned}
\tag{5.5}
$$

The group-generated and group-propagated carries for several groups can now be used to generate group carry-ins in a manner similar to single-bit carry-ins in Equation (5.4). A combinatorial circuit implementing these equations is available as a separate and standard IC. This IC is called a *carry-look-ahead generator*, and its use is illustrated in the following example.

Example 5.2

For $n = 16$ there are four groups, with outputs $G_0^*, G_1^*, G_2^*, G_3^*$ and P_0^*, P_1^*, P_2^*, P_3^*. These serve as inputs to a carry-look-ahead generator, whose

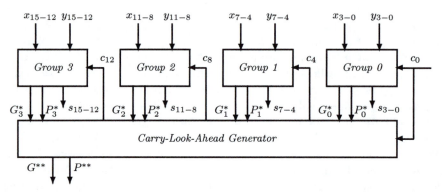

FIGURE 5.2   A 16-bit two-level carry-look-ahead adder. (The notation x_{3-0} means x_3, x_2, x_1, x_0.)

outputs are denoted by c_4, c_8, and c_{12}, satisfying

$$
\begin{aligned}
c_4 &= G_0^* + c_0 P_0^*, \\
c_8 &= G_1^* + G_0^* P_1^* + c_0 P_0^* P_1^*, \\
c_{12} &= G_2^* + G_1^* P_2^* + G_0^* P_1^* P_2^* + c_0 P_0^* P_1^* P_2^*.
\end{aligned}
\tag{5.6}
$$

● A 16-bit adder with four groups, each with internal carry-look-ahead and an additional carry-look-ahead generator, are depicted in Figure 5.2. The operation of this adder consists of the following four steps:

1. All groups generate in parallel bit-carry-generate, G_i, and bit-carry-propagate, P_i.

2. All groups generate in parallel group-carry-generate, G_i^*, and group-carry-propagate, P_i^*.

3. The carry-look-ahead generator produces the carries c_4, c_8, and c_{12} into the groups.

4. The groups calculate their individual sum bits (in parallel) with internal carry-look-ahead. In other words, they first generate the internal carries according to Equation (5.4) and then the sum bits.

The minimum time delay associated with steps (1)–(4) (assuming a minimum number of gate levels in all circuits) is $1\Delta_G$ for step 1, $2\Delta_G$ for step 2, $2\Delta_G$ for step 3, and $4\Delta_G$ for step 4. Thus, the total addition time is $9\Delta_G$ instead of $11\Delta_G$, which is the addition time if the external carry-look-ahead generator is not used and the carry ripples among the groups. These calculations yield only theoretical estimates for the addition time. In practice, one has to use the typical delays associated with the particular integrated circuits employed in order to calculate the addition time more accurately (see any integrated circuit databook).   □

As shown in Figure 5.2, the carry-look-ahead generator produces two additional outputs, G^{**} and P^{**}, whose Boolean equations are similar to those in Equation (5.5). These new outputs are called *section-carry generate* and *section-carry propagate*, respectively, where a section, in this case, is a set of four groups and consists of 16 bits. As before, the number of groups in a section is commonly set at four because of implementation-related considerations, and not because of any limitation of the underlying algorithm.

If the number of bits to be added is larger than 16, say, 64, we may use either four circuits, each similar to the one shown in Figure 5.2, with a ripple-carry between adjacent sections, or use another level of carry-look-ahead, and achieve a faster execution of addition. This is exactly the same circuit as above, accepting the four pairs of section-carry-generate and section-carry-propagate, and producing the carries c_{16}, c_{32}, and c_{48}.

As the number of bits, n, increases, more levels of carry-look-ahead generators can be added in order to speed up the addition. The required number of levels (for maximum speed up) approaches $\log_b n$, where b is the *blocking factor*; i.e., the number of bits in a group, the number of groups in a section, and so on. The blocking factor is 4 in the conventional implementation depicted in Figure 5.2. The overall addition time of a carry-look-ahead adder is therefore proportional to $\log_b n$.

5.3 CONDITIONAL SUM ADDERS

Another scheme for fast addition that provides a logarithmic speed-up is the conditional sum adder [29]. The principle behind this scheme is to generate two sets of outputs for a given group of operand bits, say, k bits. Each set includes k sum bits and an outgoing carry. One set assumes that the eventual incoming carry will be zero, while the other assumes that it will be one. Once the incoming carry is known, we need only to select the correct set of outputs (out of the two sets) without waiting for the carry to further propagate through the k positions (see Figure 5.3). Clearly, we should not apply this idea to all n operand bits at the beginning of the add operation, since we will then have to wait until the carry propagates through all n positions before making the selection. We need, therefore, to divide the given n bits into smaller groups and apply the above idea to each of them separately. In this way, the serial carry-propagation inside the separate groups can be done in parallel, reducing the overall execution time. These groups can, in turn, be further divided into subgroups, for which the carry-propagation time is even smaller. The outputs of the subgroups are then combined to generate the output of the groups.

A natural division of the n operand bits would be into two groups of size $n/2$ bits each. Each one of these can be further divided into two groups of size $n/4$ bits each. This process can, in principle, be continued until a group of size

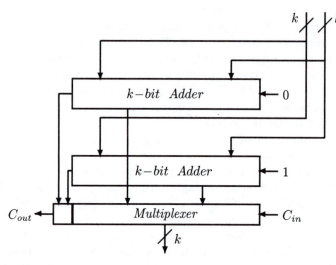

FIGURE 5.3   Selecting the correct set of sum bits and carry-out.

1 is reached if n is an integer power of 2. In this case, $\log_2 n$ steps are needed in the process, where in step 1 we deal with single bit positions, in step 2 pairs of bits are handled, and so on. Notice, however, that a given group does not necessarily have to be divided into equal-sized subgroups. Thus, the conditional sum scheme can be applied even if the number of bits is not a power of 2.

Example 5.3

In this example we illustrate the way groups containing single bits are combined into pairs of bits. We use here the following notation: s_i^0 denotes the sum bit at position i under the assumption that the incoming carry into the currently considered group is 0; s_i^1 is defined similarly, and so are the outgoing carries (from the group), c_{i+1}^0 and c_{i+1}^1. We will first consider two adjacent bit positions, and, in step 1, each constitutes a separate group:

i	7	6	
x_i	1	0	
y_i	0	0	
s_i^0	1	0	Assuming incoming carry = 0
c_{i+1}^0	0	0	
s_i^1	0	1	Assuming incoming carry = 1
c_{i+1}^1	1	0	

In step 2 the two bit positions are combined (using data selectors) into one group of size 2:

$i, i-1$	7,6	
x_i, x_{i-1}	10	
y_i, y_{i-1}	00	
s_i^0, s_{i-1}^0	10	Assuming incoming carry $= 0$
c_{i+1}^0	0	
s_i^1, s_{i-1}^1	11	Assuming incoming carry $= 1$
c_{i+1}^1	0	

Note that the carry-out from position 6 becomes an internal (to the group) carry and consequently, we can select the appropriate set of outputs for position 7. □

Example 5.4

We now apply the conditional sum method to the addition of two 8-bit operands (Figure 5.4). The process has $\log_2 8 = 3$ steps.

Notice that the forced carry (which equals 0 in this example) is available at the beginning of the operation. Therefore, only one set of outputs needs to be generated for the rightmost group at each step. □

	i	7	6	5	4	3	2	1	0
	x_i	1	0	1	1	0	1	1	0
	y_i	0	0	1	0	1	1	0	1
Step 1	s_i^0	1	0	0	1	1	0	1	1
	c_{i+1}^0	0	0	1	0	0	1	0	0
	s_i^1	0	1	1	0	0	1	0	
	c_{i+1}^1	1	0	1	1	1	1	1	
Step 2	s_i^0	1	0	0	1	0	0	1	1
	c_{i+1}^0	0		1		1		0	
	s_i^1	1	1	1	0	0	1		
	c_{i+1}^1	0		1		1			
Step 3	s_i^0	1	1	0	1	0	0	1	1
	c_{i+1}^0	0				1			
	s_i^1	1	1	1	0				
	c_{i+1}^1	0							
Result		1	1	1	0	0	0	1	1

FIGURE 5.4   Conditional sum addition of two 8-bit numbers.

A variation of the conditional sum adder is the carry-select adder. As in the conditional sum adder, the n bits are divided into groups (but not necessarily of the same size), and each group generates two sets of sum bits and an outgoing carry bit. The incoming carry selects one of these two sets. Unlike the conditional sum adder, each group is not further divided into smaller subgroups. The carry-select adder and some of its variations are described in Section 5.8.

Comparing the conditional sum and the carry-look-ahead schemes, we see that the two have about the same speed. The design of a conditional sum adder is, however, less modular than that of a carry-look-ahead adder, and this is the main reason for the much higher popularity of the latter. The next section includes a more general discussion on the optimality of algorithms for addition and their implementations.

5.4 OPTIMALITY OF ALGORITHMS AND THEIR IMPLEMENTATIONS

Numerous algorithms for fast addition, as well as other arithmetic operations, have been developed and implemented since the early days of digital computers, and newer ones are still being proposed. The main reason for the continuing research and development of new algorithms for the basic arithmetic operations is the rapid change in the technology that is used to implement them. An algorithm that can be optimally implemented in one technology is not necessarily the best in a different technology. Consequently, designers need to continuously reevaluate the available algorithms for a certain arithmetic operation and their suitability to the current technology.

In addition to the dependence on the technology employed for the implementation of the algorithm, the performance of a given algorithm is heavily affected by the unique features of the algorithm itself and/or the number system used to represent the operands and results. Thus, many studies have been performed that compare various algorithms in an effort to determine which will perform better, preferably independently of the technology used for the implementation. More importantly, the objective of some studies was to find the limit (bound) on the performance of any algorithm in executing a given arithmetic operation.

The execution time of addition, being highly dependent on the way carries are propagated, can reach its minimum in algorithms, which avoid the propagation of carries altogether, or incur a very limited carry-propagation. Therefore, number systems that are characterized by almost carry-free addition, can provide "optimal" algorithms for addition. One such number system is the residue number system (described in Chapter 11); another is the *SD* number system (described in Chapter 2). One should, however, be aware of the fact that these number systems are not frequently used in practice. Consequently, in order to take advantage of these fast algorithms for addition, we need to perform conversions between number systems. These conversions may turn out to be as complex as (or even more complex than) the addition itself and unless a fast conversion algorithm is available, this approach is of limited practical value.

A theoretical model to determine a lower bound on the speed of addition has been proposed by Winograd [35], Spira [30], and others. This is an idealized model, whose purpose is the derivation of a bound independent of the implementation technology. It assumes that the circuit for addition is realized using only one type of gate, the (f, r) gate, where r is the radix of the number system used and f is the *fan-in* of the gate; i.e., the maximum number of inputs to the gate. All (f, r) gates are assumed to be capable of computing any r-valued function of f (or less) arguments in exactly the same time period. This fixed time period is defined as the unit delay, and the computation time of the adder circuit is measured in these unit delays.

Since an (f, r) gate can compute any function of f arguments, we need to find out only how many such gates are required and how many circuit levels are needed in order to properly connect these gates. Thus, a circuit for adding two radix-r operands with n digits each has $2n$ inputs and produces $n + 1$ outputs. Consider the output that requires all $2n$ inputs for its calculation. These $2n$ inputs can be reduced to a smaller number of arguments by using $\lceil 2n/f \rceil$ such (f, r) gates (where the ceiling $\lceil x \rceil$ of a number x is the smallest integer that is larger than or equal to x). These gates belong to the same logic level and therefore operate in parallel. The resulting number of intermediate arguments is $\lceil 2n/f \rceil$, and this number can be further reduced through a second level of (f, r) gates. A schematic diagram of the resulting circuit is depicted in Figure 5.5. The total number of levels in such a tree constructed of (f, r) gates is at least $\lceil \log_f 2n \rceil$. Note that the indicated number of (f, r) gates at each level is only

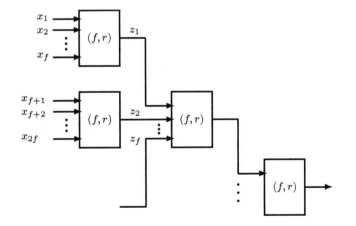

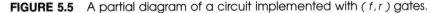

FIGURE 5.5   A partial diagram of a circuit implemented with (f, r) gates.

a lower bound, since it assumes that no argument is needed as input to more than one (f, r) gate. The resulting number of levels is consequently also a lower bound. Therefore, the lower bound on the time to perform addition is

$$T_{add} \geq \lceil \log_f 2n \rceil \tag{5.7}$$

measured in units of (f, r) gate delay.

There are several assumptions underlying this model that make it an idealized model. First, only the fan-in limitation is taken into account, while the *fan-out* constraint is ignored. The fan-out of a gate is the ability of its output to drive a number of inputs to similar gates in the next level. In practice, the fan-out of a gate is constrained. Second and more important, the model assumes that any r-valued function of f arguments can be calculated by a single (f, r) gate in one unit delay. In practice, only a small number of such functions can be implemented by a single gate requiring the smallest possible delay. Many functions may require either a more complex gate (with a longer delay) or need to be implemented using several simple gates organized in two or more levels.

The bound in Equation (5.7) assumes that there is at least one output digit that depends on *all* $2n$ input digits. If the addition technique for the particular number system employed is such that not all $2n$ digits are needed to determine any output digit, then a lower value for the bound can be established. Instead of having trees with $\lceil \log_f 2n \rceil$ levels, smaller trees (with fewer inputs) can be used. This occurs if a carry cannot propagate from the least-significant position to the most-significant position, since such a long carry propagation implies that the most-significant output digit is dependent on all $2n$ inputs. For example, if only x_i, y_i, x_{i-1} and y_{i-1} are needed to determine the sum digit s_i, then

$$T_{add} \geq \lceil \log_f 4 \rceil.$$

If the conventional binary number system is used, a carry can propagate through all n positions and $\lceil \log_f 2n \rceil$ is still a lower bound on the addition time. The two addition algorithms described in the previous sections (namely, the carry-look-ahead and the conditional sum) have an execution time that is proportional to $\log n$ and can, in theory, approach the above bound.

However, when comparing two or more algorithms with the same theoretical bound for the execution time, some objective function related to the cost of implementation should also be taken into account. The type of the additional objective function used depends on the technology employed. For example, when discrete gates are used to implement the circuit, the number of such gates may serve as an objective function measuring the implementation cost. The number of gates along the critical (longest) path (in other words, the number of circuit levels) determines the execution time of the algorithm. If full custom VLSI technology is used, then the exact number of gates has very limited effect on the implementation cost. Instead, regularity of the design and length of interconnections are considerably more important, since they affect both the silicon area used by the adder and the design time. The two factors (i.e., implementation cost and speed) do not necessarily achieve their minimum value in the same design. Thus, a tradeoff between these two might have to be found.

If performance is more important than implementation cost, then the carry-look-ahead adder is very attractive. Still, the implementation cost can be reduced especially when full custom VLSI is employed, and, as a result, regularity of the design and size of the required area determine the implementation cost. This can be achieved by taking advantage of the available degree of freedom in the design; namely, the blocking factor. The blocking factor is always bounded by the fan-in constraint. In some cases it might also be bounded by additional constraints, such as the number of pins. However, the highest possible value for the blocking factor is not necessarily the best choice, and there is a need to reevaluate its effect on execution time and implementation cost. For example, a blocking factor of 2 results in a very regular layout of binary trees with up to $\log_2 n$ levels, requiring a total area of approximate size $n \cdot \log_2 n$ [2]. Further details on the structure of the carry-look-ahead tree with a blocking factor of 2 are provided in the next section.

If lower implementation cost is required, then the carry-look-ahead scheme might be inappropriate. The ripple-carry method can be used together with some speed-up techniques, which depend on the chosen technology. One such technique is the Manchester adder [14], whose schematic diagram is shown in Figure 5.6. This diagram includes switches that can be realized using pass transistors or similar devices in various technologies. The three switches for unit

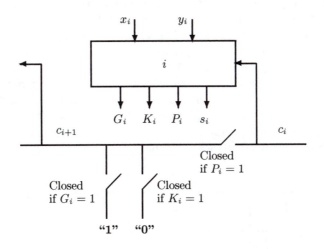

FIGURE 5.6   A Manchester adder.

i are controlled by the signals P_i, G_i, and K_i, where $P_i = x_i \oplus y_i$ is the *carry-propagate* signal, $G_i = x_i y_i$ is the *carry-generate* signal, and $K_i = \overline{x_i}\ \overline{y_i}$ is the *carry-kill* signal. These three signals are defined so that one, and only one, of the corresponding switches is closed at any time. Thus, the equation $P_i = x_i \oplus y_i$ is used instead of $P_i = x_i + y_i$ as before. If $G_i = 1$, an outgoing carry is generated irrespective of the incoming carry. If $K_i = 1$, any incoming carry is "killed" and not allowed to continue its propagation. If $P_i = 1$ however, an incoming carry is allowed to propagate.

All switches in units 0 through $n-1$ are set simultaneously, and, as a result, the propagating carry experiences only a single switch delay per stage. The number of carry-propagate switches that can be cascaded is limited in practice, and this limit depends on the specific technology employed. Thus, there is a need to partition the n units into groups and insert separating devices (buffers) between them. In theory, the execution time of this adder is linearly proportional to the number of bits, n. However, the ratio between its execution time and that of another adder (e.g., the carry-look-ahead adder) depends on the particular technology. In any technology that is employed to realize the Manchester adder, the implementation cost, measured in size of area and/or regularity of design, is expected to be lower than that of a carry-look-ahead adder [25].

5.5 CARRY-LOOK-AHEAD ADDITION REVISITED

In this section we derive the equations for carry-look-ahead addition in a more general way. This will allow us to consider various implementations of the carry-look-ahead adder rather than being restricted to a predetermined blocking factor. It will also provide a general framework for deriving expressions for other techniques for fast addition, including carry-select and carry-skip adders.

We first introduce the following notation. Let $P_{i:j}$ and $G_{i:j}$ denote the group-propagated carry and the group-generated carry functions, respectively, for the group of bit positions i, $i-1$, $\cdots$, j (with $i \geq j$), as shown in Figure 5.7.

$P_{i:j}$ equals 1 when an incoming carry into the least significant position j, c_j, is allowed to propagate through all $i - j + 1$ bit positions. $G_{i:j}$ equals 1 when a carry is generated in at least one of the bit positions from j to i inclusive and propagates to bit position $i + 1$, i.e., the outgoing carry c_{i+1} equals 1. These definitions generalize those in Equation (5.5) and include as a special case

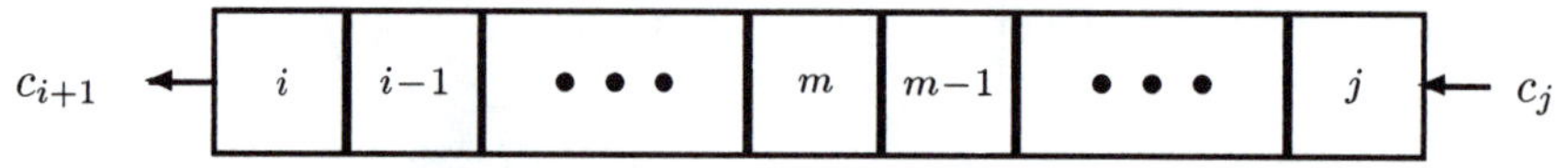

FIGURE 5.7   A group consisting of i - j + 1 bit positions ($i \geq j$).

the single bit-position carry propagate and generate functions P_i and G_i. The two group-carry functions can be calculated recursively using the two Boolean equations

$$P_{i:j} = \begin{cases} P_i & \text{if } i = j \\ P_i \cdot P_{i-1:j} & \text{if } i > j; \end{cases} \tag{5.8}$$

$$G_{i:j} = \begin{cases} G_i & \text{if } i = j \\ G_i + P_i \cdot G_{i-1:j} & \text{if } i > j. \end{cases} \tag{5.9}$$

Note that the notations $P_{i:i}$ and P_i (and similarly $G_{i:i}$ and G_i) are equivalent. The recursive Equations (5.8) and (5.9) can be further generalized to

$$\begin{aligned} P_{i:j} &= P_{i:m} \cdot P_{m-1:j}, \\ G_{i:j} &= G_{i:m} + P_{i:m} \cdot G_{m-1:j}, \end{aligned} \tag{5.10}$$

where $i \geq m \geq j + 1$. This is the same generalization that was employed in Section 5.2 to derive the section-carry propagate and generate functions, P^{**} and G^{**}. It can be formally proved by induction on m.

Instead of dealing with each of the group-carry functions separately we introduce the pair $(P_{i:j}, G_{i:j})$ and define a new Boolean operator, called the fundamental carry operator and denoted by $\circ$, as follows [2]:

$$(P, G) \circ (\tilde{P}, \tilde{G}) = (P \cdot \tilde{P}, G + P \cdot \tilde{G}) \tag{5.11}$$

Using this fundamental carry operator we can rewrite the recursive Equation (5.10) as

$$(P_{i:j}, G_{i:j}) = (P_{i:m}, G_{i:m}) \circ (P_{m-1:j}, G_{m-1:j}), \tag{5.12}$$

where $i \geq m \geq j + 1$. The fundamental carry operation has been shown to be associative [2], i.e., for $i \geq m > v > j$ (or, more accurately, $i \geq m \geq v+1 \geq j+2$)

$$\begin{aligned} &((P_{i:m}, G_{i:m}) \circ (P_{m-1:v}, G_{m-1:v})) \circ (P_{v-1:j}, G_{v-1:j}) \\ &= (P_{i:m}, G_{i:m}) \circ ((P_{m-1:v}, G_{m-1:v}) \circ (P_{v-1:j}, G_{v-1:j})). \end{aligned} \tag{5.13}$$

From its definition in Equation (5.11) the fundamental carry operation is an idempotent operation

$$(P, G) \circ (P, G) = (P \cdot P, G + P \cdot G) = (P, G) \tag{5.14}$$

and consequently, the most general form of Equation (5.12) is

$$(P_{i:j}, G_{i:j}) = (P_{i:m}, G_{i:m}) \circ (P_{v:j}, G_{v:j}), \tag{5.15}$$

where $i \geq m$ and $v \geq j$ but v is not necessarily equal $m - 1$; it is only required that $v \geq m - 1$.

The above equations indicate how the (propagate and generate) group carries $P_{i:j}$ and $G_{i:j}$ can be calculated from subgroup carries where the two or more subgroups are of arbitrary size and these subgroups may even overlap. Eventually, we would like to use these group and subgroup carries in order to calculate the individual bit carries c_{i+1}, c_i, $\cdots$, c_{j+1}, and sum outputs s_i, s_{i-1}, $\cdots$, s_j. To this end we must take into account the "external" carry c_j (see Figure 5.7). For the mth bit position, $i \geq m \geq j$, we have

$$c_m = G_{m-1:j} + P_{m-1:j} \cdot c_j \tag{5.16}$$

which can also be rewritten as

$$(P_{m-1:j}, G_{m-1:j}) \circ (1, c_j),$$

and if $P_m = x_m \oplus y_m$ then

$$s_m = c_m \oplus P_m. \tag{5.17}$$

However, if $P_m = x_m + y_m$ then $s_m = c_m \oplus (x_m \oplus y_m)$. An alternative way to deal with the incoming carry into the group, c_j, is to modify the equation for G_j from $x_j y_j$ to $x_j y_j + P_j \cdot c_j$. Then, the equation for c_m becomes

$$c_m = G_{m-1:j}. \tag{5.18}$$

The later "forces" the carry c_j to propagate through the group while the former allows it to "skip" the group. We will further elaborate on this in the discussion of carry-skip adders.

Equations (5.15)–(5.18) can be used to derive various implementations of adders including ripple-carry, carry-look-ahead, carry-select, carry-skip and other. A 5-bit ripple-carry adder corresponds to the case where all subgroups consist of a single bit position and the computation starts at position 0, proceeds to position 1 and so on:

$$(P_4, G_4) \circ \{(P_3, G_3) \circ ((P_2, G_2) \circ [(P_1, G_1) \circ \{(P_0, G_0) \circ (1, c_0)\}])\} \tag{5.19}$$

A 16-bit carry-look-ahead adder with four groups of size 4 (i.e., blocking factor of 4) and a ripple-carry among groups corresponds to the following expression:

$$(P_{15:12}, G_{15:12}) \circ \{(P_{11:8}, G_{11:8}) \circ [(P_{7:4}, G_{7:4}) \circ \{(P_{3:0}, G_{3:0}) \circ (1, c_0)\}]\} \tag{5.20}$$

We next introduce a variant of a carry-look-ahead adder that was proposed in [2]. This variant uses a blocking factor of 2 resulting in a very regular layout of a binary tree with $\log_2 n$ levels, requiring a total area of approximate size $n \cdot \log_2 n$. To illustrate the design of the adder, consider the calculation of c_{16}, the incoming carry at stage 16 in a 17-bit (or more) adder and suppose that $G_0 = x_0 y_0 + P_0 \cdot c_0$.

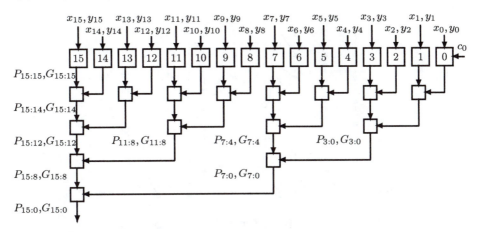

FIGURE 5.8 A tree structure for calculating c_{16} (each line, except c_0, represents two signals that are either x_m and y_m or $P_{v:m}$ and $G_{v:m}$).

The tree structure for calculating c_{16} is depicted in Figure 5.8. The part of the tree structure that generates $(P_{7:0}, G_{7:0})$ corresponds to the expression

$$
\begin{aligned}
(P_{7:0}, G_{7:0}) &= (P_{7:4}, G_{7:4}) \circ (P_{3:0}, G_{3:0}) \\
&= \{(P_{7:6}, G_{7:6}) \circ (P_{5:4}, G_{5:4})\} \circ \{(P_{3:2}, G_{3:2}) \circ (P_{1:0}, G_{1:0})\} \\
&= \{[(P_7, G_7) \circ (P_6, G_6)] \circ [(P_5, G_5) \circ (P_4, G_4)]\} \\
&\quad \circ \{[(P_3, G_3) \circ (P_2, G_2)] \circ [(P_1, G_1) \circ (P_0, G_0)]\}
\end{aligned}
\qquad (5.21)
$$

All the circuits in the second through the fifth levels in Figure 5.8 implement the fundamental carry operation and are therefore identical. c_{16} is equal to $G_{15:0}$ (compare to Equation (5.18)) and, since $P_m = x_m \oplus y_m$, we can calculate the sum bit s_{16} using $s_{16} = c_{16} \oplus P_{16}$. The tree structure in Figure 5.8 also generates the carries c_2, c_4 and c_8. The carry bits for the remaining bit positions can be calculated through extra subtree structures that can be added to the binary tree shown in Figure 5.8. Once all the carries are known, the corresponding sum bits can be computed using Equation (5.17).

In the above design the blocking factor always equals 2. However, the blocking factor does not have to be the same for all levels of the carry generation tree. Different values of the blocking factor may lead to a more efficient use of space and/or shorter interconnections [22].

5.6 PREFIX ADDERS

The adder shown in Figure 5.8 may be viewed as a parallel prefix circuit. A parallel prefix circuit is a combinational circuit with n inputs $x_1, x_2, \cdots, x_n$ pro-

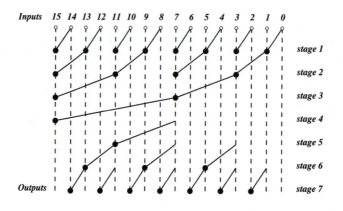

FIGURE 5.9   The Brent-Kung (2) parallel prefix graph.

ducing the outputs $x_1, x_2 \circ x_1, \cdots, x_n \circ x_{n-1} \circ \cdots \circ x_1$, where $\circ$ is an associative binary operation. The first stage of the adder in Figure 5.8 generates the individual P_i and G_i signals. The remaining stages constitute the parallel prefix circuit with the fundamental carry operation serving as the $\circ$ associative binary operation. This part of the adder tree can be designed in many different ways. The particular way this part is implemented within the 16-bit Brent-Kung adder [2] in Figure 5.8 is shown in Figure 5.9. The bullets implement the fundamental carry operation while the empty circles at the top generate the individual P_i and G_i signals. Note that Figure 5.8 which generates $G_{15:0}$ shows only the top four stages of the complete parallel prefix graph in Figure 5.9 which uses seven stages to also generate all the intermediate carries $G_{i:0}$ $(i = 14, 13, \cdots, 1)$.

   The number of stages and consequently, the total delay of the adder, can be reduced by modifying the structure of the parallel prefix graph. The minimum number of stages for a parallel prefix graph is $\log_2 n$ which for $n = 16$ is equal to 4, while the number of stages in a Brent-Kung parallel prefix graph is $2 \log_2 n - 1$. One way to implement a four-stage parallel prefix graph has been proposed in [17] and is shown in Figure 5.10. Note that unlike Figure 5.9, the Ladner-Fischer

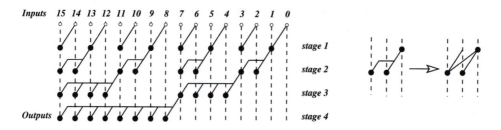

FIGURE 5.10   The Ladner-Fischer (17) parallel prefix graph.

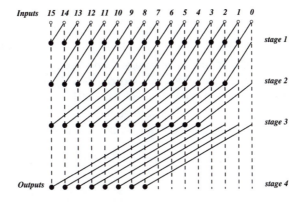

FIGURE 5.11 The Kogge-Stone (16) parallel prefix graph.

adder employs fundamental carry operators with a fan-in value higher than 2, i.e., the blocking factor varies from 2 to $n/2$. Such an implementation also implies a fan-out of up to $n/2$ requiring buffers which add to the overall delay.

Another parallel prefix graph which also uses only $\log_2 n$ stages but has lower fan-in and fan-out requirements, has been proposed in [16] and is shown in Figure 5.11. This adder has still a higher number of lateral wires with a longer span than the Brent-Kung adder and such wires usually require some buffering, resulting in additional delay. Several other variants of parallel prefix graphs have been proposed (e.g., [15]) illustrating that in general, smaller adder delay can be achieved in exchange for higher overall area and/or power. Compromises between simplicity of wiring and overall delay have also been suggested. For example, a hybrid design combining stages from the Brent-Kung and Kogge-Stone adders was proposed in [12] and is shown in Figure 5.12. It has five rather than four stages with the middle three resembling the Kogge-Stone structure, but its wires have a shorter span than those in the Kogge-Stone adder.

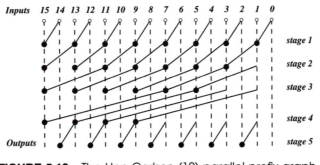

FIGURE 5.12 The Han-Carlson (12) parallel prefix graph.

5.7 LING ADDERS

Ling adders [19] are a variation of the carry-look-ahead adders. They use a simpler version of the group-generated carry signal and thus provide an opportunity to reduce the associated delay. We will see the principle behind Ling adders through a simple example. Suppose that we start with a carry-look-ahead adder with groups of size 2, i.e., we produce the signals $G_{1:0}$, $P_{1:0}$, $G_{3:2}$, $P_{3:2}$ and so on. The outgoing carry for position 3 can be expressed as

$$c_4 = G_{3:0} = G_{3:2} + P_{3:2}G_{1:0} \qquad (5.22)$$

where

$$G_{3:2} = G_3 + P_3G_2, \quad G_{1:0} = G_1 + P_1G_0, \text{ and } P_{3:2} = P_3P_2.$$

We either assume that $c_0 = 0$ or set $G_0 = x_0y_0 + P_0c_0$. Expressing $G_{3:0}$ in terms of the individual carry generate and propagate signals we obtain

$$G_{3:0} = G_3 + P_3G_2 + P_3P_2(G_1 + P_1G_0). \qquad (5.23)$$

Since $G_3P_3 = G_3$ all terms in the above equation have P_3 as a common factor, and we can therefore rewrite the expression for $G_{3:0}$ as

$$G_{3:0} = P_3H_{3:0},$$

where $H_{3:0}$ is defined as

$$H_{3:0} = H_{3:2} + P_{2:1}H_{1:0}, \qquad (5.24)$$

and

$$H_{3:2} = G_3 + G_2, \quad H_{1:0} = G_1 + G_0.$$

Note that $P_{2:1}$ is used in (5.24) in contrast to $P_{3:2}$ which is used in (5.22). Equation (5.24) defines H as an alternative to the carry generate G and shows that H can be calculated in a similar manner. However, H does not have a simple interpretation like G. On the other hand, H is simpler to calculate. For example, expressing $H_{3:0}$ in terms of the individual bit signals yields

$$H_{3:0} = G_3 + G_2 + P_2P_1(G_1 + G_0)$$

which can be simplified to

$$H_{3:0} = G_3 + G_2 + P_2G_1 + P_2P_1G_0, \qquad (5.25)$$

while Equation (5.23) can be rewritten as

$$G_{3:0} = G_3 + P_3G_2 + P_3P_2G_1 + P_3P_2P_1G_0. \qquad (5.26)$$

The maximum fan-in in (5.25) is smaller than that in (5.26) leading to simpler and usually faster circuits. Other variations of the expression for the group-generated carry G have corresponding variations for H. For example, for the equation $G_{3:0} = G_3 + P_3 G_{2:0}$, the corresponding equation for H is $H_{3:0} = G_3 + T_2 H_{2:0}$ where $T_2 = x_2 + y_2$. A more general expression for H is $H_{i:0} = G_i + T_{i-1} H_{i-1:0}$ where $T_{i-1} = x_{i-1} + y_{i-1}$.

The calculation of the sum bits in a Ling adder is slightly more involved than that for the carry-look-ahead. To illustrate this calculation consider s_3:

$$
\begin{aligned}
s_3 &= c_3 \oplus (x_3 \oplus y_3) = (P_2 H_{2:0}) \oplus (x_3 \oplus y_3) \\
&= \overline{H_{2:0}}(x_3 \oplus y_3) + H_{2:0}(P_2 \oplus (x_3 \oplus y_3))
\end{aligned}
\tag{5.27}
$$

The calculation of $H_{2:0}$ is faster than that of c_3 reducing the delay associated with generating s_3.

Three other variation of the carry-look-ahead adder which have properties similar to those of Ling's adder have been presented in [7]. An implementation of a 32-bit Ling adder is described in [8]. A muliplexor-based implementation of Ling adders is presented in [26].

5.8 CARRY-SELECT ADDERS

In a carry-select adder the n bits are divided into nonoverlapping groups of possibly different lengths. The underlying strategy is similar to that of the conditional-sum adder described in Section 5.3. Each group generates two sets of sum bits and an outgoing carry. One set assumes that the incoming carry into the group is 0, the other assumes that it is 1. When the incoming carry into the group is assigned its final value it selects one of the two sets as is shown in Figure 5.3. Figure 5.13 is a more detailed version of Figure 5.3 depicting the lth group which consists of k bit positions starting with bit position j and ending with bit position i where $i = j + k - 1$.

The outputs of the group are the sum bits s_i, s_{i-1}, $\cdots$, s_j and the lth group outgoing carry c_{i+1}. The corresponding Boolean equations are

$$
s_m = s_m^0 \cdot \overline{c_j} + s_m^1 \cdot c_j ; \qquad m = j, j+1, \cdots, i,
\tag{5.28}
$$

and

$$
c_{i+1} = c_{i+1}^0 \cdot \overline{c_j} + c_{i+1}^1 \cdot c_j,
\tag{5.29}
$$

where s_m^0 is the mth sum bit under the condition that the incoming carry into the lth group is 0. This is the same notation that we have used for the conditional-sum adder. The notations s_m^1, c_{i+1}^0 and c_{i+1}^1 are defined similarly.

The two separate sets of outputs can be calculated in a ripple-carry manner. Thus, for bit position m we calculate c_m^0 and c_m^1 from $G_{m-1:j}^0$ and $G_{m-1:j}^1$,

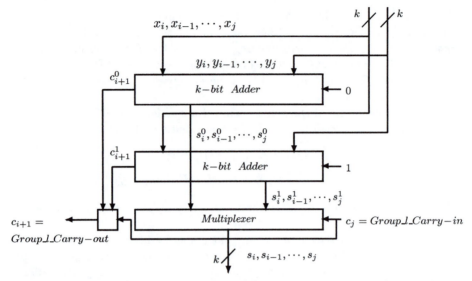

FIGURE 5.13 The *l* th group, consisting of the *k* bit positions *j*, *j* + 1, $\cdots$, *l*, in a carry-select adder.

respectively, which in turn, are calculated from (see Equation (5.19))

$$(P_{m-1:j}, G^0_{m-1:j}) = (P_{m-1}, G_{m-1}) \circ (P_{m-2}, G_{m-2}) \circ \cdots \circ (P_j, G_j) \qquad (5.30)$$

and

$$(P_{m-1:j}, G^1_{m-1:j}) = (P_{m-1:j}, G^0_{m-1:j}) \circ (1, 1) = (P_{m-1:j}, G^0_{m-1:j} + P_{m-1:j}).$$
$$(5.31)$$

Note that $P_{m-1:j}$ has no superscript since it is independent of the incoming carry. Once the individual bit carries have been calculated the corresponding sum bits are

$$s^0_m = c^0_m \oplus P_m \qquad \text{and} \qquad s^1_m = c^1_m \oplus P_m.$$

Since c^0_{i+1} implies c^1_{i+1} (i.e., if $c^0_{i+1} = 1$ then c^1_{j+1} must equal 1) we can simplify Equation (5.29) to read

$$c_{i+1} = c^0_{i+1} + c^1_{i+1} \cdot c_j. \qquad (5.32)$$

The sizes of the separate groups can be either different (e.g., [33]), or they can all be equal to k (e.g., [1]), with possibly one group of size smaller than k. In the first case the size of the lth group is chosen so as to equalize the delay of the ripple-carry within the group and the delay of the carry-select chain from group 1 to group l. The above two delays depend on the technology employed and the exact implementation. If, for example, we assume a simple two-level gate implementation for the multiplexing circuit corresponding to Equation (5.32) then the delay associated with the carry-select chain through the preceding $l-1$ groups is $(l-1) \cdot 2\Delta_G$ where Δ_G is the delay of a single gate. The delay of the ripple-carry chain through the k_l bit positions in the lth group, when again we assume a simple two-level gate implementation, is $k_l \cdot 2\Delta_G$. Equalizing these two delays results in

$$k_l = l-1 \quad \text{with} \quad k_l \geq 1 \; ; \quad l = 1, 2, \cdots, L \qquad (5.33)$$

where L is the number of groups as shown below.

In other words, the group lengths should follow the simple arithmetic progression 1, 1, 2, 3, $\cdots$, and the total number of bits, n, must satisfy

$$1 + L(L-1)/2 \geq n, \qquad (5.34)$$

and consequently

$$L(L-1) \geq 2(n-1). \qquad (5.35)$$

As a result, the size of the largest group and the execution time of the carry-select adder are of the order of $\sqrt{n}$. For example, for $n = 32$, based on Equation (5.35) nine groups are required. One possible choice for their sizes is 1, 1, 2, 3, 4, 5, 6, 7 and 3. The total carry propagation time, under the assumption of two-level gate implementation, is $18 \cdot \Delta_G$, instead of $62\Delta_G$ for the ripple-carry adder.

   If the lengths of all L groups are equal, the carry-select chain (i.e., generating the $group\_l\_Carry-out$ from the $group\_l\_Carry-in$, see Figure 5.13) does not have to be necessarily of the ripple-carry type. Instead, a single or even multiple-level carry-look-ahead can be employed [1].

   Compared to the ripple-carry adder, the carry-select adder requires a duplicate carry-chain logic and additional carry-select logic. However, this logic circuit overhead can be reduced by observing that the two separate carry-chains in Figure 5.13 and the multiplexing circuitry can be combined to yield a simpler implementation [33]. In such an implementation each bit position includes the following logic circuitry

$$G_m = x_m y_m \; ; \quad P_m = x_m \oplus y_m \; ; \quad P_{m:0} = P_m \cdot P_{m-1:0},$$

and

$$c_m^0 = G_m + P_m \cdot c_{m-1}^0 \; ; \quad s_m = P_m \oplus (c_{m-1}^0 + c_j \cdot P_{m-1:0}). \qquad (5.36)$$

The proof of these equations is left to the reader as an exercise. Other variations of the carry-select adder have been proposed and implemented with some of these described in Section 5.10.

5.9 CARRY-SKIP ADDERS

A carry-skip adder reduces the time needed to propagate the carry by skipping over groups of consecutive adder stages. As such, the carry-skip adder generalizes the idea behind the Manchester adder described in Section 5.4. The carry-skip adder illustrates the dependence of the "optimal" algorithm for addition on the available technology. Although the carry-skip algorithm has been known for many years, it has become popular only recently. In VLSI technology the carry-skip adder is comparable in speed to the carry look-ahead technique (for commonly used word lengths but not necessarily in the asymptotic sense) but it requires less chip area and consumes less power.

The carry-skip adder is based on the following observation. The carry propagation process can skip any adder stage for which $x_m \neq y_m$ (or in other words, $P_m = x_m \oplus y_m = 1$). Several consecutive stages can be skipped if all satisfy $x_m \neq y_m$. Thus, an adder consisting of n stages is divided into groups of consecutive stages with a simple ripple-carry scheme used in each group. Every group also generates a group-carry-propagate signal that equals 1 if all stages internal to the group satisfy $P_m = 1$. This signal can be used to allow an incoming carry into the group to "skip" all the stages within the group and generate a group-carry-out. Let a particular group, say, group l, consist of the k bit positions $j, j + 1, \cdots, j + k - 1$ as shown in Figure 5.14. Based on Equation (5.16) the Boolean expression for *Group_l_Carry-out* is

$$Group\_l\_Carry\text{-}out = G_{i:j} + P_{i:j} \cdot Group\_l\_Carry\text{-}in$$

where $G_{i:j}$ equals 1 when a carry is generated internal to the group and is allowed to propagate through all the remaining bit positions including i. $P_{i:j}$ equals 1

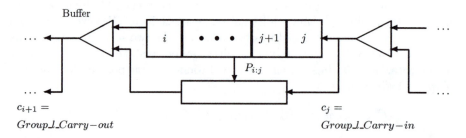

FIGURE 5.14 The *l* th group consisting of bit positions *j, j* + 1, $\cdots$, *i* in a carry-skip adder.

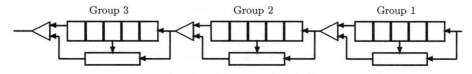

FIGURE 5.15   A 15-bit carry-skip adder.

when the $i - j + 1$ bit positions allow the incoming carry c_j to propagate to the next bit position, $i + 1$. The buffers shown in the figure realize the OR operation in the above Boolean expression.

Figure 5.15 depicts a 15-bit carry-skip adder consisting of three groups, each of size 5. Notice that the signals $P_{i:j}$ for all groups can be generated simultaneously allowing a fast skip of groups which satisfy $P_{i:j} = 1$.

We wish to determine the optimal size of the group, k. This optimal size depends on the ratio between the carry-ripple time through a single stage, denoted by t_r, and the time it takes to skip a group of size k, denoted by $t_s(k)$. The latter is, for most implementations, independent of k.

Assume first that all groups are of the same size k, and, for simplicity, assume further that n/k is an integer. The group size k should be selected so that the time for the longest carry-propagation chain is minimized. The longest carry-propagation chain occurs when a carry is generated in stage 0 and then propagates all the way to stage $n - 1$. This means that the carry will ripple through stages $1, 2, \cdots, k - 1$ within group 1, skip groups $2, 3, \cdots, (n/k - 1)$, then ripple through group n/k. The overall carry-propagation time is in this case

$$T_{carry} = (k - 1) \cdot t_r + t_b + (n/k - 2) \cdot (t_s + t_b) + (k - 1) \cdot t_r, \qquad (5.37)$$

where t_b is the delay associated with the buffer (which implements the OR operation) between two groups, as shown in Figure 5.15. If, for example, a straightforward two-level gate implementation is employed for both the ripple-carry circuit and the carry-skip circuit, then t_r would equal $t_s + t_b = 2\Delta_G$, yielding

$$T_{carry} = (4k + 2n/k - 7) \cdot \Delta_G.$$

Differentiating T_{carry} with respect to k and equating the derivative to 0 results in

$$k_{opt.} = \sqrt{n/2}.$$

As for the carry-select adder, the group size and the carry propagation time are proportional to $\sqrt{n}$.

For example, for $n = 32$, eight groups of size $k_{opt.} = \sqrt{16} = 4$ will provide the best design, with $T_{opt.} = 25\Delta_G$, instead of $62\Delta_G$ for the ripple-carry adder.

Revisiting the previous analysis, one should realize that further speed-up can be achieved if we make the size of the first and last groups even smaller than the fixed size k, and in this way reduce the ripple-carry delay through these groups. Also, we may increase the size of the center groups, since the skip time is usually independent of group size. Another way to reduce T_{carry} is to design a second level of skip circuitry that would allow skipping two or more groups in one step. Additional levels can also be envisioned.

The idea of using unequal group sizes has already been suggested in the past [18]. Only recently have several algorithms been developed for deriving the optimal group sizes for different technologies and implementations (i.e., different values of the ratio $(t_s + t_b)/t_r$).

We will now formulate the problem and illustrate its solution through several examples. Note first that, unlike the simple analysis for the equal-sized group case above, we cannot restrict ourselves to the analysis of the worst case for carry propagation. This may lead to the trivial conclusion that the first and last groups should consist of a single stage, while all remaining $n-2$ stages should constitute a single center group. In this design, a carry generated at the beginning of the center group may ripple through all the other $n-3$ stages, becoming the worst case. We therefore need to consider all possible carry-propagation chains that may start at an arbitrary bit position a (for which $x_a = y_a$) and stop at the next position b that also satisfies $x_b = y_b$, where a new carry-propagation chain (independent of the previous one) may start.

Let $k_1, k_2, \cdots, k_L$ denote the size of the L different groups starting at position 0. Then

$$\sum_{i=1}^{L} k_i = n.$$

In the most general case, a carry-propagation chain starts at some position within group u, ends at some position within group v, and skips the groups $u + 1, u + 2, \cdots, v - 1$. In the worst case, the carry will be generated in the first position within group u, and will stop in the last position within group v. The overall carry-propagation time, denoted by $T_{carry}(u, v)$, is

$$T_{carry}(u, v) = (k_u - 1) \cdot t_r + t_b + \sum_{l=u+1}^{v-1} (t_s(k_l) + t_b) + (k_v - 1) \cdot t_r. \qquad (5.38)$$

The set of group sizes $k_1, k_2, \cdots, k_L$ should be selected so that the longest carry-propagation chain is minimized:

$$minimize \left[\max_{1 \leq u \leq v \leq L} T_{carry}(u, v) \right]$$

To solve this optimization problem, the size of the groups, as well as the number of groups, L, must be determined. Algorithms for solving this problem have been developed, relying on either geometrical interpretations (e.g., [13]) or dynamic programming [4].

Example 5.5

The optimal organization for a 32-bit carry-skip adder with a single level of carry-skip has been derived in several different ways. This optimal organization includes $L = 10$ groups with sizes $k_1, k_2, \cdots, k_{10} = 1, 2, 3, 4, 5, 6, 5, 3,$ $2, 1$ for $t_s + t_b = t_r$ yielding $T_{carry} \leq 9 \cdot t_r$ [13]. If $t_r = 2\Delta_G$, then $T_{carry} \leq 18\Delta_G$, instead of $25\Delta_G$, as in the equal-size group case. The reader can verify that any two bit positions in any two groups u and v, $(1 \leq u \leq v \leq 10)$, satisfy $T_{carry}(u, v) \leq 9 \cdot t_r$.   □

The similarities between the carry-skip and carry-select adders and their carry propagation times should not come as a surprise. Although the strategies behind the two schemes sound different, the equations relating the group-carry-out with the group-carry-in are, in both cases, variations of the same basic Equation (5.16). Only the details of the implementation vary, in particular the calculation of the sum bits. Even this difference is reduced when the multiplexing circuitry is merged into the summation logic according to Equation (5.36).

5.10 HYBRID ADDERS

Hybrid adders are adders which use a combination of two or more of the previously described methods for addition. A common approach to the design of hybrid adders is to choose one method for carry propagation and another method for sum calculation. The two hybrid adders presented in this section combine some variation of a carry-select adder for calculating the sum and a modified Manchester adder for carry propagation. Both divide the operands into groups of equal size—8 bits each.

The first hybrid adder [20] employs the carry-select method for calculating the sum for each group of 8 bits separately as shown in Figure 5.16. The group carry-in signal that selects one out of the two sets of sum bits is not generated in a ripple-carry manner as shown in Figure 5.13. Instead, the carries into the 8-bit groups are generated by a carry-look-ahead tree as proposed in [1]. In the case of a 64-bit adder these are c_8, c_{16}, c_{24}, c_{32}, c_{40}, c_{48} and c_{56} (see Figure 5.16).

The structure of a carry-look-ahead tree for generating these carries would be similar but not necessarily identical to that shown in Figure 5.8. The differences between such structures stem from variations in the blocking factor at each level of the tree and the exact implementation of a module for calculating the fundamental carry operator. If we restrict ourselves to a fixed blocking factor the natural choices for groups of size 8 bits include 2 (as in Figure 5.8), 4 or 8.

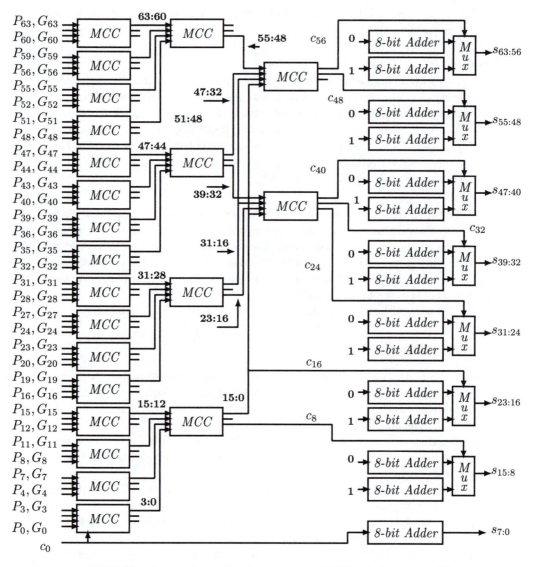

FIGURE 5.16   A schematic diagram of a 64-bit hybrid adder (20).

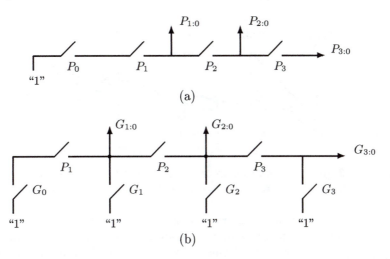

(a)

(b)

FIGURE 5.17   A Manchester carry module for calculating a group propagate (circuit a) and generate (circuit b) for a group of size 4.

The first choice results in the largest number of levels in the tree while the last one results in complex modules for the fundamental carry operator with a high delay. A blocking factor of 4 represents a reasonable compromise and has been selected in [20]. A Manchester carry propagate/generate module (MCC in Figure 5.16) with a blocking factor of four is depicted in Figure 5.17.

In the most general case the Manchester carry module in Figure 5.17 accepts four pairs of inputs: $(P_{i_1:i_0}, G_{i_1:i_0})$, $(P_{j_1:j_0}, G_{j_1:j_0})$, $(P_{k_1:k_0}, G_{k_1:k_0})$ and $(P_{l_1:l_0}, G_{l_1:l_0})$ where $i_1 \geq i_0$, $j_1 \geq j_0$, $k_1 \geq k_0$ and $l_1 \geq l_0$. It produces three pairs of outputs: $(P_{j_1:i_0}, G_{j_1:i_0})$, $(P_{k_1:i_0}, G_{k_1:i_0})$ and $(P_{l_1:i_0}, G_{l_1:i_0})$ under the conditions $i_1 \geq j_0 - 1$, $j_1 \geq k_0 - 1$ and $k_1 \geq l_0 - 1$. These conditions allow overlap among the input subgroups following Equation (5.15). A schematic diagram showing the operation of the carry module is depicted in Figure 5.18.

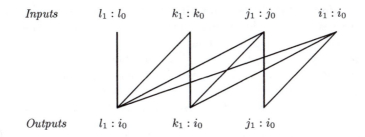

FIGURE 5.18   A schematic diagram describing the operation of the Manchester carry module in Figure 5.17 in the general case.

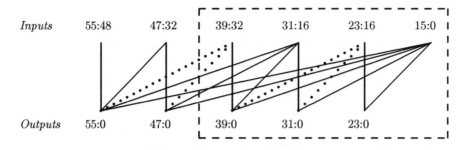

FIGURE 5.19 The available inputs and required outputs at the third level of the carry-look-ahead tree. (A dotted line represents an optional dependence.)

The first level of the carry-look-ahead tree for a 64-bit adder includes 14 Manchester carry modules and calculates $(P_{3:0}, G_{3:0})$, $(P_{7:4}, G_{7:4})$, $\cdots$, $(P_{55:52}, G_{55:52})$, i.e., only the outputs $P_{3:0}$ and $G_{3:0}$ in Figure 5.17 are utilized. In the second level of the carry-look-ahead tree each Manchester carry module generates two pairs of outputs, corresponding to $(P_{3:0}, G_{3:0})$ and $(P_{1:0}, G_{1:0})$ in Figure 5.17. Thus, the second level of modules provides the values $(P_{7:0}, G_{7:0})$, $(P_{15:0}, G_{15:0})$, $(P_{23:16}, G_{23:16})$, $(P_{31:16}, G_{31:16})$, $(P_{39:32}, G_{39:32})$, $(P_{47:32}, G_{47:32})$, and $(P_{55:48}, G_{55:48})$ (see Figure 5.16). This level generates the carries c_8 and c_{16}, which equal $G_{7:0}$ and $G_{15:0}$, respectively, since c_0 is already incorporated into the module that generates $(P_{3:0}, G_{3:0})$ in the first level [20] (see Equation (5.18)).

The available inputs, required outputs and the dependence among them in the third level of the carry-look-ahead tree are shown in Figure 5.19. Clearly two Manchester carry modules are sufficient to produce the required outputs in Figure 5.19. One such module can implement the relationships included in the dashed box in this figure. This module will generate the carries c_{24}, c_{32} and c_{40}. A second module can produce the two remaining pairs of required outputs with the inputs corresponding to 55:48, 47:32, 31:16 and 15:0. This module will generate the carries c_{48} and c_{56}. Notice in particular, that the first module described by the dashed box in Figure 5.19 must implement the two dotted lines from 23:16 in the figure since 23:16 is required for generating 23:0. The above described implementation of a 64-bit adder combining the carry-select scheme for generating the sum bits and a Manchester-based carry-look-ahead tree for generating the carries into the groups is not unique and does not necessarily minimize the overall execution time. Alternate implementations including variable size of the carry-select groups and of the Manchester carry modules at the different levels of the tree may prove to achieve a lower execution time.

The 64-bit adder described in [6] divides the 64 bits into two sets of size 32 bits. Each set of 32 bits is, in turn, further divided into four groups of size 8 bits. For every group of eight bits two sets of conditional sum outputs are

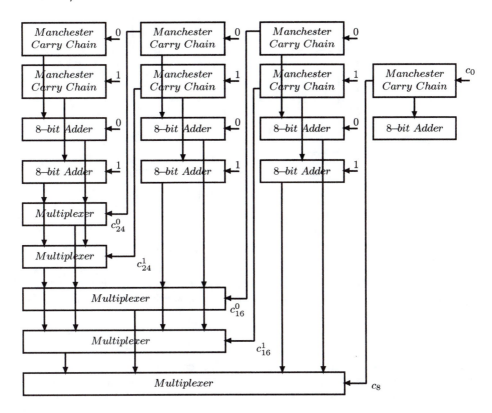

FIGURE 5.20   A schematic diagram of a 32-bit hybrid adder.

generated separately. The two most significant groups are then combined into a single larger group of size 16. This larger group is further combined with the next group of size 8 to form an even larger group of size 24 bits and so on following the principle of conditional-sum addition. However, the way the input carries for the basic 8-bit groups are generated is completely different from the method described in Section 5.3. A schematic diagram depicting the low-order half of the 64-bit adder is shown in Figure 5.20. The Manchester carry chain unit in this figure generates the P_m, G_m and K_m signals for the individual bits and computes the c_{out}^0 and c_{out}^1 outputs for the assumed incoming carry of 0 and 1, respectively. These conditional carry-out signals control the multiplexers and determine which conditional sum bits are forwarded to the next level of multiplexing. The two sets of dual multiplexers (of size 8 and 16 bits) and the single regular multiplexer of size 24 bits are implemented using the circuits shown in Figure 5.21. The high-order half of the 64-bit adder has a structure similar to that in Figure 5.20. The

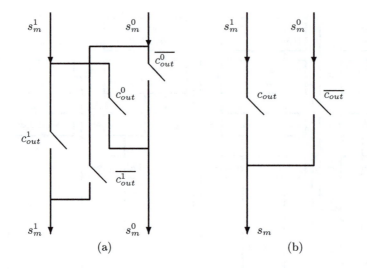

FIGURE 5.21 The basic circuit for a dual multiplexer (a) and a single multiplexer (b). The signals c_{out}^0 and c_{out}^1 are generated by the Manchester carry unit of the preceding group of 8 bits.

main difference is that the incoming carry, c_{32}, is calculated by a separate carry-look-ahead circuit whose inputs are the conditional carry-out signals generated by the four Manchester carry units in Figure 5.20. This allows the operation of the high-order half of the 64-bit adder to overlap the operation of the low-order half. In summary, this adder combines variants of three different techniques for fast addition: Manchester carry generation, carry-select and conditional-sum.

Other designs of hybrid adders can be envisioned, combining variations of the basic methods for fast addition, possibly implementing for example, groups with unequal number of bits. One such adder, a Manchester adder with variable (group size) carry-skip, has been proposed and analyzed in [3]. The "optimality" of hybrid adders is highly dependent on the available technology and its particular delay parameters.

5.11 CARRY-SAVE ADDERS

When three or more operands are to be added simultaneously (e.g., in multiplication) using two-operand adders, the time-consuming carry-propagation must be repeated several times. If the number of operands is k, then carries have to propagate $(k - 1)$ times. Several techniques for multiple operand addition that attempt to lower the carry-propagation penalty have been proposed and implemented. The technique that is most commonly used is *carry-save addition*. In

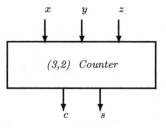

FIGURE 5.22 A (3,2) counter.

carry-save addition, we let the carry propagate only in the last step, while in all the other steps we generate a partial sum and a sequence of carries separately. Thus, the basic carry-save adder (CSA) accepts three n-bit operands and generates two n-bit results, an n-bit partial sum, and an n-bit carry. A second CSA accepts these two bit-sequences and another input operand, and generates a new partial sum and carry. A CSA is therefore, capable of reducing the number of operands to be added from 3 to 2, without any carry propagation.

A carry-save adder may be implemented in several different ways. In the simplest implementation, the basic element of the carry-save adder is a full adder with three inputs, x, y, and z, whose arithmetic operation can be described by

$$x + y + z = 2c + s, \tag{5.39}$$

where s and c are the sum and carry outputs, respectively. Their values are

$$s = (x + y + z) \bmod 2 \quad \text{and} \quad c = \frac{(x + y + z) - s}{2}. \tag{5.40}$$

The outputs are the weighted binary representation of the number of 1's in the inputs. We therefore call the FA a $(3, 2)$ counter, as shown in Figure 5.22. An n-bit CSA consists of n (3,2) counters operating in parallel with no carry links interconnecting them.

A carry-save adder for four 4-bit operands X, Y, Z, and W, is shown in Figure 5.23. The upper two levels are 4-bit CSAs, while the third level is a 4-bit carry-propagating adder (CPA). The latter is a ripple-carry adder, but may be replaced by a carry-look-ahead adder or any other fast CPA. One should note that partial sum bits and carry bits are interconnected to guarantee that only bits having the same weight are added by any (3,2) counter.

In order to add the k operands X_1, X_2, $\cdots$, X_k we need $(k-2)$ CSA units and one CPA. If the CSAs are arranged in a cascade, as in Figure 5.23, then the time to add the k operands is

$$(k - 2) \cdot T_{CSA} + T_{CPA},$$

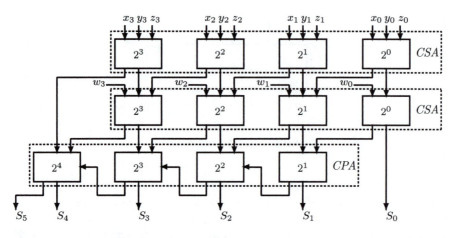

FIGURE 5.23   A carry-save adder for four operands.

where T_{CPA} is the operation time of a CPA and T_{CSA} is the operation time of a CSA, which equals the delay of a full adder, Δ_{FA}. The latter is at least $2 \cdot \Delta_G$, where Δ_G is the delay of a single gate. Note that the final result may reach a length of $n + \lceil log_2 k \rceil$ bits, since the sum of k operands, of size n bits each, can be as large as $(2^n - 1)k$.

A better way to organize the CSAs, and reduce the operation time, is in the form of a tree commonly called a Wallace tree [34]. A six-operand Wallace tree is illustrated in Figure 5.24. The left arrows on the carry outputs of the CSAs indicate that these outputs have to be shifted to the left before being added to the sum bits, as shown in Figure 5.23. In this tree, the number of operands is reduced by a factor of 2/3 at each level. Thus,

$$k \cdot (\frac{2}{3})^l \leq 2$$

where l is the number of levels required. Consequently,

$$\text{Number of levels} \approx \frac{log\ (k/2)}{log\ (3/2)}. \tag{5.41}$$

Equation (5.41) provides only an estimate of the number of levels, since at each level the number of operands must be an integer. Thus, if N_i is the number of operands at level i, then the number of operands at the level $(i+1)$ above can be at most $\lfloor N_i \cdot 3/2 \rfloor$ (where the floor $\lfloor x \rfloor$ of a number x is the largest integer that is smaller than or equal to x). The number of operands at the bottom level (i.e., level 0) is 2, so that the maximum number of operands at level 1 is 3 and the maximum number of operands at level 2 is $\lfloor 9/2 \rfloor = 4$. The resulting

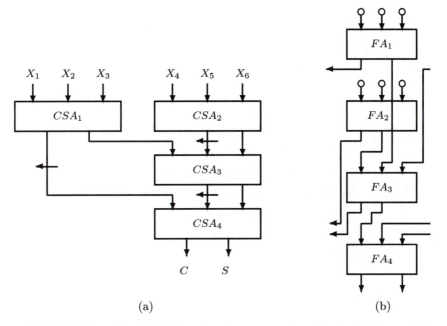

FIGURE 5.24   (a) A CSA tree for six operands. (b) An implementation of a
6-input bit-slice of the tree in (a).

sequence of numbers is 2,3,4,6,9,13,19,28, etc. Starting with five operands, we
still need three levels as we do for six operands. The entries in Table 5.1 were
generated using similar arguments. This table shows the exact number of levels
required for up to 63 operands.

Example 5.6

For $k = 12$, five levels are needed, resulting in a delay of $5 \cdot T_{CSA}$, instead
of $10 \cdot T_{CSA}$, which is the delay for a linear cascade of 10 CSAs.   □

Examining Table 5.1, we may note that the most economical implementation (in terms of number of levels) is achieved when the number of operands
is an element of the series 3,4,6,9,13,19,28, $\cdots$. Thus, for a given number of
operands, say, k, which is not an element of this series, we need to use only
enough CSAs to reduce k to the closest (and smaller than k) element in the
above series. For example, for $k = 27$, we may use 8 CSAs (with 24 inputs)
rather than 9 CSAs, in the top level, so that the number of operands in the next
level will be $8 \cdot 2 + 3 = 19$, which is an element of the series. The remaining part
of the tree will have its operands follow the series.

Number of operands	Number of levels
3	1
4	2
$5 \leq k \leq 6$	3
$7 \leq k \leq 9$	4
$10 \leq k \leq 13$	5
$14 \leq k \leq 19$	6
$20 \leq k \leq 28$	7
$29 \leq k \leq 42$	8
$43 \leq k \leq 63$	9

TABLE 5.1   The number of levels in a CSA tree for k operands.

The idea of using a (3,2) counter to form multi-operand adders can be extended to a (7,3) counter, whose three outputs represent the number of 1's in its seven inputs. Another example is the (15,4) counter or, in general, any (k,m) counter where k and m satisfy

$$2^m - 1 \geq k \quad \text{or} \quad m \geq \lceil log_2(k+1) \rceil.$$

A (7,3) counter, for example, can be implemented using (3,2) counters as shown in Figure 5.25, where intermediate results are added according to their weight. However, this implementation requires four (3,2) counters arranged in three levels and therefore provides no speed-up compared to an implementation based on (3,2) counters. A (7,3) counter can also be implemented directly as a multi-level circuit that may have a smaller overall delay depending on the particular technology employed [21]. Since the number of interconnections that a circuit requires greatly affects its silicon area, a (7,3) counter is preferrable to a (3,2) counter. A (7,3) has ten connections and removes four bits while a (3,2) counter has five connections and removes only one bit. Another implementation of the (7,3) counter is through a ROM of size $2^7 \times 3 = 128 \times 3$ bits. The access time of this ROM is unlikely to be smaller than the delay associated with the implementation in Figure 5.25. However, a speed-up may be achieved if a ROM implementation is used for a (k,m) counter with higher values of k and m.

When several (7,3) counters (in parallel) are used to add seven operands, we obtain three results, and a second level of (3,2) counters is needed to reduce these to two results (sum and carry) to be added by a CPA. A similar situation arises when (15,4) or more complex counters are used, generating more than two results and consequently requiring a second level of counters. In some cases, the additional level of counters can be combined with the first level of counters, resulting in a more convenient implementation.

In what follows, we show how the (7,3) counter can be combined with a (3,2) counter. We call the combined counter a (7;2) compressor. In general,

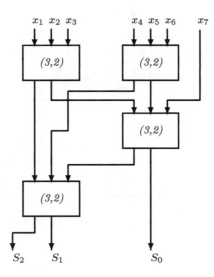

FIGURE 5.25   A (7,3) counter using (3,2) counters.

a $(k; m)$ compressor is a variant of a counter with k primary input, all of the same weight, say 2^i, and m primary outputs of weights 2^i, 2^{i+1}, $\cdots$, 2^{i+m-1} [9]. In addition, the compressor has several incoming carries, all of weight 2^i, from previous compressors, and several outgoing carries of weights 2^{i+1} and up. The 6-input bit-slice shown in Figure 5.24(b) is a trivial example of a (6;2) compressor where all outgoing carries have the same weight 2^{i+1} and the number of these carries equals the number of incoming carries and is also equal, in general, to $k - 3$.

A straightforward implementation of a (7;2) compressor is shown in Figure 5.26, where the bottom right (3,2) counter is the additional (3,2) counter, while the remaining four (3,2) counters constitute the ordinary (7,3) counter that is depicted in Figure 5.25. A (7;2) compressor in column i has seven primary inputs of weight 2^i and two carry inputs from column $(i - 1)$ and $(i - 2)$. It generates two primary outputs, denoted by $S2^i$ and $S2^{i+1}$, reflecting their weights, and two outgoing carries $C2^{i+1}$ and $C2^{i+2}$, to columns $(i + 1)$ and $(i + 2)$, respectively. Note that the input carries to the (7;2) compressor in Figure 5.26 do not participate in the generation of the two output carries in order to avoid a slow carry-propagation chain. Also notice that this compressor is not a (9,4) counter since it has two outputs $(S2^{i+1}$ and $C2^{i+1})$ with the same weight. The implementation depicted in Figure 5.26 does not offer any speedup. A different, possibly multilevel logic, implementation may yield a smaller overall delay as long as the generation of the outgoing carries remains independent of the incoming carries.

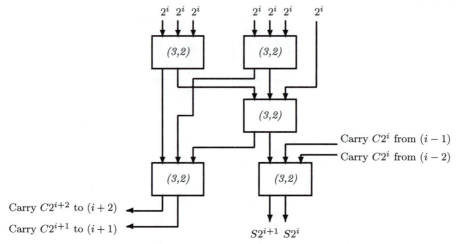

FIGURE 5.26   A (7;2) compressor with two incoming carries and two outgoing carries, in bit position i.

All the previously described counters are single-column counters. For multi-operand addition we can generalize these single-column counters into multiple-column counters. We define a generalized parallel counter as a counter that adds l input columns and produces an m-bit output [31]. The notation we use for such a counter is

$$(k_{l-1}, k_{l-2}, \cdots, k_0, m)$$

where k_i is the number of input bits in the i−th column with weight 2^i. Clearly, a (k, m) counter is a special case of this generalized counter. The number of outputs m must satisfy

$$2^m - 1 \geq \sum_{i=0}^{l-1} k_i 2^i. \tag{5.42}$$

If all l columns have the same height k (i.e., $k_0 = k_1 = ... = k_{l-1} = k$), then the inequality that has to be satisfied is

$$2^m - 1 \geq k \cdot (2^l - 1). \tag{5.43}$$

A simple example of these counters is the (5,5,4) counter shown in Figure 5.27. For this counter, $k = 5$, $l = 2$ and $m = 4$, and inequality (5.43) turns into an equality, implying that all 16 combinations of the output bits are useful. (5,5,4) counters can be used to reduce five operands (of any length) to two results that can then be added with a CPA. The length of operands will determine the number of (5,5,4) counters in parallel.

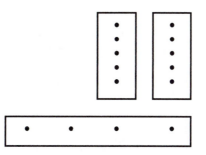

FIGURE 5.27   A (5,5,4) counter. The dots represent input or output bits.

A reasonable way of implementing generalized counters is by using ROMs. For example, the (5,5,4) counter shown in Figure 5.27 can be realized with a $2^{(5+5)} \times 4$ ROM (i.e., 1024×4).

The (5,5,4) counters conveniently reduce the input operands to two intermediate results, requiring only one CPA to produce the final sum. In the general case, a string of $(k_0 = k, ..., k_{l-1} = k, m)$ counters may generate more than two intermediate results, requiring additional reduction before a CPA can be used. To find out the number of intermediate results generated by a set of $(k, k, \cdots, k, m)$ counters, consider the following. A set of $(k, k, \cdots, k, m)$ counters, with l columns each, produces m-bit outputs at intervals of l bits. Any column has at most $\lceil \frac{m}{l} \rceil$ output bits. Thus, k operands can be reduced to $s = \lceil \frac{m}{l} \rceil$ operands. If $s = 2$, a single CPA can generate the final sum. Otherwise, further reduction, from s to 2, is needed.

Example 5.7

If the number of bits per column in a two-column counter (k, k, m) is increased beyond 5, then $m \geq 5$ and as a result, $s = \lceil \frac{m}{2} \rceil > 2$. For example, if $k = 7$, the inequality that must be satisfied is $2^m - 1 \geq 7 \cdot 3 = 21$, and therefore $m = 5$. A set of (7,7,5) counters will generate $s = 3$ operands, and consequently another set of (3,2) counters is needed in order to reduce the number of operands to 2.   □

The hardware complexity of a carry-save adder for a large number of operands might be prohibitive, independent of the particular type of parallel counters employed. One way to reduce the hardware complexity is to design a smaller carry-save tree and use it iteratively. The n operands are divided into $\lceil n/j \rceil$ groups of j operands each, and a tree for $j + 2$ operands with two feedback paths and a CPA is designed, as shown in Figure 5.28. The two feedback paths make it necessary to complete the first pass through the CSA tree before

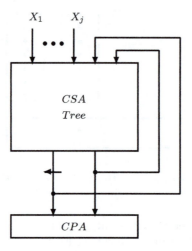

FIGURE 5.28   A CSA tree with two feedback paths and j new operands.

the second set of j operands is applied. This slows down the execution of the multiple-operand addition, since pipelining is not possible. In the next section we discuss pipelining in general and describe ways to modify the tree structure in Figure 5.28 to support pipelining.

5.12 PIPELINING OF ARITHMETIC OPERATIONS

Pipelining is a very well known technique for accelerating the execution of successive identical operations. Instead of designing a circuit capable of executing a single operation on one set of operands at a time, we design one that is partitioned into several subcircuits that can operate independently on consecutive sets of operands. This way the executions of several successive operations overlap, and the rate at which results are produced is considerably higher than that of a nonpipelined design.

To allow pipelining, the algorithm is divided into several steps, and a suitable circuit is designed for each of these steps. The separate circuits, which are called *pipeline stages*, must be allowed to operate independently on different sets of operands. To achieve this goal, storage elements (latches) must be added between adjacent stages, so that when a stage works on one set of operands, the preceding stage can work on the next set of operands.

An example of a pipeline for addition (or any other two-operand operation) consisting of three stages is depicted in Figure 5.29. Here, the addition of the two operands X and Y is performed in three steps. The latches between stage 1 and stage 2 store the intermediate results of step 1, which are then used by

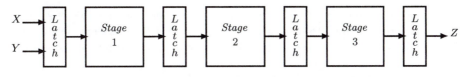

FIGURE 5.29   A three-stage pipeline.

stage 2 to execute step 2 of the algorithm, while stage 1 can start the execution of step 1 on the next set of operands X and Y.

A common way to illustrate the way a pipeline operates is through a timing diagram like the one in Figure 5.30, which shows the exact timing of four successive additions with operands X_1 & Y_1, X_2 & Y_2, X_3 & Y_3, and X_4 & Y_4 producing the results Z_1, Z_2, Z_3, and Z_4, respectively.

Let τ_i denote the execution time of stage i and let τ_l denote the time needed to store new data into a latch. In general, the delays associated with the different pipeline stages are not identical, and faster stages must wait until the slowest stage completes its task before they can all switch to the next task. Therefore, the time interval between two successive final results being produced by the pipeline is

$$\tau = \max_{1 \le i \le k} \{\tau_i\} + \tau_l \tag{5.44}$$

where k is the number of stages in the general case. The time interval τ is also called the *pipeline period*, and $1/\tau$ is called the *pipeline rate* or *bandwidth*.

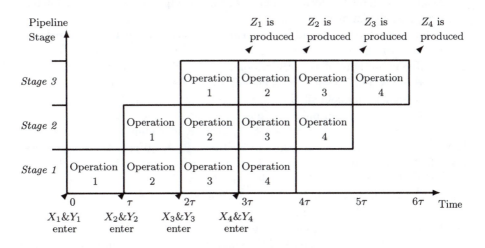

FIGURE 5.30   A timing diagram of a three stage pipeline.

The clock signal synchronizing the pipeline's operation must be set so that the clock period is equal to or larger than τ. Figure 5.30 shows the case where the clock period equals τ. Here, after a latency of $3 \cdot \tau$, new results are produced at the rate of $1/\tau$.

An important design decision is the partitioning of the given algorithm into steps that will be executed by the separate stages of the pipeline. These steps should preferably be defined so that they have similar execution times, since the pipeline rate is determined by the execution time of the slowest step. The number of steps must then be determined. As this number increases, the pipeline period decreases, but the number of latches goes up (increasing the cost of implementation), and so does the latency of the pipeline. The latency is the time elapsed until the first result is produced. This is especially important when only a single pass through the pipeline is required (e.g., addition of only one pair of operands). Thus, there is a tradeoff between the latency and implementation cost on one hand and the pipeline rate on the other hand. The extra delay due to the latches, τ_l, can be lowered by using special circuits like the Earl latch [11].

5.12.1 Pipelining of Adders

The relative simplicity of two-operand adders usually does not justify their implementation as pipelines. However, in special-purpose designs, when many successive additions are needed, such implementations are justifiable. The conditional-sum adder can be easily implemented as a pipeline. One way of doing this is to have $\log_2 n$ stages corresponding to the $\log_2 n$ steps in the conditional-sum algorithm. This allows us to overlap the execution of up to $\log_2 n$ additions. However, the required number of latches may be excessive. Two (or even more) steps can be combined to form a single stage in the pipeline reducing the latches' overhead and the latency.

The carry-look-ahead adder, described in Section 5.2, cannot be pipelined, since some carry signals must propagate backward (see, e.g., Figure 5.2). However, different designs of the carry-look-ahead adder, following the approach described in Section 5.5, can be pipelined. Here, the final carries and the carry-propagate signals (implemented as $P_i = x_i \oplus y_i$ rather than $P_i = x_i + y_i$) can be used to calculate the sum bits, eliminating the need for feedback connections.

Clearly, pipelining is more beneficial in the case of multiple-operand adders, like the carry-save adders described in Section 5.9. Modifying the implementation of CSA trees (see for example, Figure 5.24) to form a pipeline is straightforward and requires only the addition of latches. These can be added at each level of the tree if maximum bandwidth is desired, or two (or more) levels of the tree can be combined to form a single stage of the pipeline, reducing the overall number of latches and the pipeline latency.

If the hardware complexity of the CSA tree for a large number of operands is prohibitive, a partial tree like the one shown in Figure 5.28 can be designed.

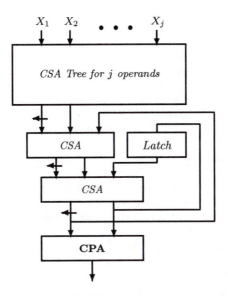

FIGURE 5.31   A CSA tree allowing overlap between iterations.

However, as pointed out earlier, the two feedback connections prevent pipelining. This can be rectified by modifying these feedback connections. Instead of connecting the two intermediate results of the CSA tree to its inputs, we can connect them to the bottom level of the CSA tree structure, as shown in Figure 5.31. The modified structure now consists of a smaller tree with j inputs at the top, two separate CSAs, and a set of latches at the bottom. The two separate CSAs and latches assimilate the two intermediate results and form a pipeline stage. This way, the top CSA tree for j operands can be pipelined too, and the overall time needed to add all n operands is reduced considerably.

5.13 EXERCISES

5.1. Compare the two alternatives for the Boolean expression of the propagated carry, namely, $P_i = x_i \oplus y_i$ and $P_i = x_i + y_i$. What might be the benefits and drawbacks of each expression?

5.2. **(a)** A *carry-completion* adder [10] detects the completion of the carry propagation and generates a signal CC (carry complete), indicating that the sum bits are steady and can be used. The basic unit in this adder is a modified full adder with the same inputs x_i and y_i and output s_i. The carries are different, though; instead of a single incoming carry line c_i, there are two incoming carry lines (and, similarly, two outgoing carry lines) denoted by c_i^0 and c_i^1. $c_i^0 = 1$ if it is *already known* that the incoming carry is zero, and $c_i^1 = 1$ if it is already known that the incoming carry is one. The two outgoing carries, c_{i+1}^0 and c_{i+1}^1, are defined similarly.

Note that an adder stage with inputs $x_i = y_i = 0$ can never generate a carry-out and therefore can produce $c_{i+1}^0 = 1$ immediately, without waiting for any carry propagation. Similarly, an adder stage with inputs $x_i = y_i = 1$ will always generate a carry-out and can therefore produce $c_{i+1}^1 = 1$ immediately without waiting for any carry propagation. An adder stage with inputs $x_i y_i = 01$ or $x_i y_i = 10$ will initially set both carry-out signals to $c_{i+1}^0 = c_{i+1}^1 = 0$ and will wait until one (and exactly one) of its carry-in signals becomes 1. Only then either c_{i+1}^0 or c_{i+1}^1 is set to 1. Write Boolean equations for the three outputs s_i, c_{i+1}^0 and c_{i+1}^1 as functions of the four inputs x_i, y_i, c_i^0, and c_i^1. Define a signal $CC_i = c_{i+1}^0 + c_{i+1}^1$, explain its meaning, and show the Boolean equation for the carry-completion signal CC.

(b) It has been shown that the average length of the longest carry propagation chain when two n-bit operands are added is approximately $\log_2(5n/4)$. Estimate the average addition time and compare it to the addition time of a ripple-carry adder. Take $n = 64$ as a numerical example. What is the drawback of this carry-completion adder?

5.3. Verify that the organization with 10 groups of size $k_1, k_2, \cdots, k_{10} = 1, 2, 3, 4, 5, 6, 5, 3, 2, 1$ for a 32-bit carry-skip adder with a single level of carry-skip and $t_s + t_b = t_r$ satisfies $T_{carry} \le 9 \cdot t_r$ [13]. You may either analyze all important cases or write a program that enumerates all cases.

5.4. Estimate the carry-propagation time in a two-level carry-skip adder for 32 bits that are divided into 14 groups of size $1,2,(1,2,3),(2,3,3),(3,3,2),(3,2),2$, where every set of parentheses signifies a single group in the second level. The first five groups, of size $1,2,(1,2,3)$, are shown in Figure 5.32 [32]. Assume that $t_s + t_b = t_r$ and compare your estimate to the carry propagation delay of the single-level carry-skip adder in problem (3). Is the second level of carry-skip justified?

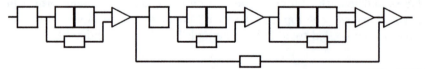

FIGURE 5.32   A two-level carry-skip adder.

5.5. Estimate the addition time of an 80-bit carry-look-ahead adder constructed of the ICs 74181 and 74182 for various adder configurations, including a ripple-carry between 74181 ICs and the maximum number of levels of 74182 ICs. Draw a block diagram for each configuration. Base your estimation on typical delays from input signals to output signals, which can be found in any IC data book.

5.6. Using a table similar to the one in Figure 5.4, show the various steps of the conditional-sum addition of the following two numbers, each consisting of 24 bits:
$x =$ 000101101100101101001111
$y =$ 001001110000011110110111

5.7. **(a)** Show an implementation of a 16-bit conditional-sum adder using full adders and data selectors. Indicate how many ICs are needed. Repeat the IC count for a 24-bit adder.
(b) Estimate the execution time of the conditional-sum adders in (a) and compare it to the execution time of carry-look-ahead adders. Base your estimation on typical delays from input signals to output signals, which can be found in any IC data book.

5.8. **(a)** Design a 32-bit conditional sum adder that starts off with groups of size 4. The two possible sets of outputs for each group of size 4 are generated using 74181 ICs with internal carry-look-ahead. Use appropriate data selectors in addition to the 74181 ICs.
(b) Estimate the execution time of the conditional-sum adder in (a) and compare it to the execution time of a carry-look-ahead adder.

5.9. Add to the tree structure in Figure 5.5 the minimum necessary circuitry to generate the carries $c_{15}, c_{14}, \cdots c_8$.

5.10. Prove the expression for estimating the number of levels in a Wallace tree for k operands.

5.11. Design a 4-bit counter capable of reducing the number of operands from 4 to 2 with a carry whose propagation is limited to one position.

5.12. A set of $n/2$ (5,5,4) counters can be used to reduce 5 operands (n-bit each) to 2 operands.
(a) What kind of counter is needed to reduce 7 operands to 2?
(b) Repeat (a) for 9 operands.
(c) Repeat (a) for k operands.

5.13. One level of (5,5,4) counters can be used to reduce 5 operands to 2 operands. What is the maximum number of operands that can be reduced to 2 when two levels of (5,5,4) counters are used?

5.14. Show an implementation of a (5,5,4) counter using (3,2) counters.

5.15. Design a (7;2) compressor with 7 inputs, 2 outputs (as in Figure 5.26) that has 4 input carries from position $(i-1)$ and generates 4 output carries to position $(i+1)$. Use (3,2) counters and make sure that no long carry-propagation chains are generated. Compare this design to the one shown in Figure 5.26 considering speed of operation and other factors.

5.16. Estimate the time needed to add n operands using the CSA tree for j new operands shown in Figure 5.28, and compare it to the time needed using the CSA tree shown in Figure 5.31 with overlap between successive iterations. Assume that n/j is an integer. Use $n = 24$ and $j = 6$ as a numerical example.

5.17. A CSA tree for 6 operands of length 32 bits each includes 4 CSA units. What should the length of each of these CSA units be?

5.18. Prove Equation (5.36) for digit position m in the reduced complexity carry-select adder.

5.19. Show how to incorporate the carry c_0 in the Manchester carry module shown in Figure 5.17.

5.20. Draw a schematic diagram like the one shown in Figure 5.17 for the 8-bit Manchester Carry Chain used in Figure 5.20. For each bit position m ($m = 0, 1, \cdots, 7$) there are three switches: P_i, G_i and K_i. Explain why is K_i needed here while it is not required in Figure 5.17.

5.14 REFERENCES

[1] O. J. BEDRIJ, "Carry-select adder," *IRE Trans. on Electron. Computers, EC-11* (June 1962), 340-346.

[2] R. P. BRENT and H. T. KUNG, "A regular layout for parallel adders," *IEEE Trans. on Computers, C-31* (March 1982), 260-264.

[3] P. K. CHAN and M. D. F. SCHLAG, "Analysis and design of CMOS Manchester adders with variable carry-skip," *IEEE Trans. on Computers, 39* (August 1990), 983-992.

[4] P. K. CHAN, M. D. F. SCHLAG, C. D. THOMBORSON and V. G. OKLOBDZIJA, "Delay optimization of carry-skip adders and block carry-look-ahead adders using multidimensional dynamic programming," *IEEE Trans. on Computers, 41* (August 1992), 920-930.

[5] L. DADDA, "Some schemes for parallel multipliers," *Alta Frequenza, 34* (March 1965), 346-356.

[6] D. W. DOBBERPUHL *et al.*, "A 200-MHz 64-b dual-issue CMOS microprocessor," *IEEE J. of Solid-State Circuits, 27* (Nov. 1992), 1555-1565.

[7] R. W. DORAN, "Variants of an Improved Carry Look-Ahead Adder," *IEEE Trans. on Computers, 37* (Sept. 1988), 1110-1113.

[8] M. J. FLYNN AND S. F. OBERMAN, *Advanced computer arithmetic design*, Wiley, New York, 2001.

[9] D. D. GAJSKI, "Parallel compressors," *IEEE Trans. on Computers, C-29* (May 1980), 393-398.

[10] B. GILCHRIST, J. POMERENCE and S. Y. WONG, "Fast carry logic for digital computers," *IRE Trans. on Electron. Computers, EC-4* (Dec. 1955), 133-136.

[11] T. G. HALLIN and M. J. FLYNN, "Pipelining of arithmetic functions," *IEEE Trans. on Computers, C-21* (August 1972), 880-886.

[12] T. HAN and D. A. CARLSON, "Fast area-efficient VLSI adders," *Proc. 8th Symp. on Computer Arithmetic,* 1987, 49-56.

[13] V. KANTABUTRA, "Designing optimum carry-skip adders," *Proc. 10th Symp. on Computer Arithmetic,* 1991, 146-153.

[14] T. KILBURN, D. B. G. EDWARDS and D. ASPINALL, "A parallel arithmetic unit using a saturated-transistor fast-carry circuit," *Proc. of IEE, Pt. B, 107* (Nov. 1960), 573-584.

[15] S. KNOWLES, "A family of adders," *Proc. 14th Symp. on Computer Arithmetic,* 1999, 30-34.

[16] P. M. KOGGE and H. S. STONE, "A parallel algorithm for the efficient solution of a general class of recurrence equations," *IEEE Trans. on Computers, C-22* (August 1973), 786-793.

[17] R. E. LADNER and M. J. FISCHER, "Parallel prefix computation," *Journal of ACM, 27* (October 1980), 831-838.

[18] M. LEHMAN and N. BURLA, "Skip techniques for high-speed carry propagation in binary arithmetic units," *IRE Trans. on Electron. Computers, EC-10* (Dec. 1962), 691-698.

[19] H. LING, "High speed binary adder," *IBM J. Res. and Devel., 25* (May 1981), 156-166.

[20] T. LYNCH and E. E. SWARTZLANDER, JR., "A spanning tree carry look-ahead adder," *IEEE Trans. on Computers, 41* (August 1992), 931-939.

[21] M. MEHTA, V. PARMAR and E. SWARTZLANDER, "High-speed multiplier design using multi-input counter and compressor circuits," *Proc. 10th Symp. on Computer Arithmetic,* 1991, 43-50.

[22] T. F. NGAI, M. J. IRWIN and S. RAWAT, "Regular, area-time efficient carry-look-ahead adders," *J. of Parallel and Distributed Computing, 3* (1986) 92-105.

[23] V. G. OKLOBDZIJA, "Design and analysis of fast carry-propagate adder under non-equal input signal arrival profile," *Proc. 28th Asilomar Conference,* (1994), 1398-1401.

[24] V. G. OKLOBDZIJA and E. R. BARNES, "On implementing addition in VLSI technology," *J. of Parallel and Distributed Computing, 5* (1988) 716-727.

[25] S. ONG and D. E. ATKINS, "A comparison of ALU structures for VLSI technology," *Proc. 6th Symp. on Computer Arithmetic,* (June 1983), 10-15.

[26] D.S. PHATAK and I. KOREN, "Intermediate variable encodings for two operand addition enabling multiplexor-based implementations," *Proc. 14th IEEE Symp. on Computer Arithmetic,* 1999, 22-29.

[27] D.S. PHATAK, T. GOFF and I. KOREN, "Constant-time addition and simultaneous format conversion based on redundant binary representations," *IEEE Trans. on Computers, 50,* (2001).

[28] S. SINGH and R. WAXMAN, "Multiple operand addition and multiplication," *IEEE Trans. on Computers, C-22* (1973) 113-120.

[29] J. SKLANSKY, "Conditional-sum addition logic," *IRE Trans., EC-9* (June 1960) 226-231.

[30] P. S. SPIRA, "Computation times of arithmetic and Boolean functions in (d, r) circuits," *IEEE Trans. on Computers, C-22* (June 1973) 552-555.

[31] W. J. STENZEL, W. J. KUBITZ and G. H. GARCIA, "A compact high-speed parallel multiplication scheme," *IEEE Trans. on Computers, C-26* (Oct. 1977) 948-957.

[32] S. TURRINI, "Optimal group distribution in carry-skip adders," *Proc. 9th Symp. on Computer Arithmetic,* 1989, 96-103.

[33] A. TYAGI, "A reduced-area scheme for carry-select adders," *IEEE Trans. on Computers, C-42* (October 1993), 1163-1170.

[34] C. S. WALLACE, "A suggestion for a fast multiplier," *IEEE Trans. on Computers, EC-13* (February 1964) 14-17.

[35] S. WINOGRAD, "On the time required to perform addition," *J. of the ACM, 12* (1965) 277-285.

6

HIGH-SPEED MULTIPLICATION

Multiplication involves two basic operations: the generation of partial products and their accumulation. Consequently, there are two ways to speed up multiplication: reduce the number of partial products or accelerate their accumulation. Clearly, a smaller number of partial products also reduces the complexity, and, as a result, reduces the time needed to accumulate the partial products.

High-speed multipliers can be classified into three general types. The first generates all partial products in parallel, and then uses a fast multi-operand adder for their accumulation. This is known as a *parallel multiplier*. The second, known as a high-speed *sequential multiplier*, generates the partial products sequentially and adds each newly generated product to the previously accumulated partial product. The third is made up of an array of identical cells that generate new partial products and accumulate them simultaneously. Thus, there are no separate circuits for partial product generation and for their accumulation. This is known as an *array multiplier*, and it tends to have a reduced execution time, at the expense of increased hardware complexity.

6.1 REDUCING THE NUMBER OF PARTIAL PRODUCTS

To reduce the number of partial products (and hence reduce the amount of hardware involved and the execution time) we may examine two or more bits of the multiplier at a time. However, this scheme requires the generation of the multiples A, $2A$, and $3A$, where A is the multiplicand, as in Chapter 3. This reduces the number of partial products to $n/2$, but each step becomes

141

more complex. Various algorithms for reducing the number of partial products without increasing the complexity of generating each partial product have been proposed. One of the first such algorithms was Booth's algorithm [3].

Booth's algorithm, as well as many other algorithms, is based on the fact that fewer partial products have to be generated for groups of consecutive zeros and ones. For a group of consecutive zeros in the multiplier there is no need to generate any new partial product. We only need to shift the previously accumulated partial product one bit position to the right for every 0 in the multiplier. For a group of, say m, consecutive 1's in the multiplier, $\cdots 0\{11\cdots 11\}0\cdots$, fewer than m new partial products can be generated. The above sequence equals the difference between the following two bit sequences, each having a single nonzero bit:

$$\cdots 0\{11\cdots 11\}0\cdots = \quad \cdots 1\{00\cdots 00\}0\cdots \quad - \quad \cdots 0\{00\cdots 01\}0\cdots$$

Using SD (signed-digit) notation, discussed in Chapter 2, the above can be written as

$$\cdots 1\{00\cdots 0\bar{1}\}0\cdots$$

For example, $\cdots 0\{1111\}0\cdots = \quad \cdots 1\{0000\}0\cdots \quad - \quad \cdots 0\{0001\}0\cdots$ $= \cdots 1\{000\bar{1}\}0\cdots$ or, in decimal notation, $15 = 16-1$. Thus, instead of generating all m partial products, we may generate only two partial products, with the second being complemented. In other words, the first partial product is added, while the second is subtracted. Note that the required number of single-bit shift-right operations is still m.

We call this operation *recoding* the multiplier in SD code. The simplest recoding scheme is the original Booth's algorithm, summarized in Table 6.1. In this algorithm, the current bit x_i and the previous bit x_{i-1} of the multiplier $x_{n-1}x_{n-2}\cdots x_1 x_0$ are examined in order to generate the ith bit y_i of the recoded multiplier $y_{n-1}y_{n-2}\cdots y_1 y_0$. The previous bit, x_{i-1}, serves here only as a reference bit. At its turn, x_{i-1} will be recoded to yield y_{i-1}, with x_{i-2} serving as a reference bit. For $i = 0$, we define the reference bit x_{-1} to be zero. A simple way of computing the recoded bit is through $y_i = x_{i-1} - x_i$.

The recoding of the multiplier bits need not be done in any predetermined order (from the most significant bit to the least significant bit or vice versa) and can even be done in parallel for all bit positions.

x_i	x_{i-1}	Operation	Comments	y_i
0	0	shift only	string of zeros	0
1	1	shift only	string of ones	0
1	0	subtract and shift	beginning of a string of ones	$\bar{1}$
0	1	add and shift	end of a string of ones	1

TABLE 6.1   Booth's algorithm.

	x_{n-1}	x_{n-2}	y_{n-1}
(1)	1	0	$\bar{1}$
(2)	1	1	0

TABLE 6.2   Recoding the sign bit in Booth's algorithm.

Example 6.1

The multiplier $0011110011(0)$ is recoded as $01000\bar{1}010\bar{1}$, requiring 4, instead of 6, add/subtract operations. The zero in parentheses is the reference bit x_{-1} for x_0.   □

When the multiplier and multiplicand are represented in two's complement, Booth's algorithm yields the correct product if the sign bit x_{n-1} participates in the process. For this sign bit, we need to decide whether to perform an add or subtract operation. However, no shift operation (of the accumulated partial product) is required, since this shift operation serves only as preparation for the next step. Clearly, the correctness of the last statement has to be verified only for negative values of X (for which $x_{n-1} = 1$). Thus, there are two cases that need to be examined and they are shown in Table 6.2. In both cases the required product is, as we have seen in Chapter 3,

$$A \cdot X = A \cdot \widetilde{X} - A \cdot x_{n-1} \cdot 2^{n-1} \quad \text{where} \quad \widetilde{X} = \sum_{j=0}^{n-2} x_j 2^j. \tag{6.1}$$

In case (1), y_{n-1} calls for the subtraction of A, which is done after the partial product has been shifted $(n-1)$ times. Hence, the necessary correction is made. In case (2), without considering the sign bit we are scanning over a string of 1's and we need to perform an addition for position $(n-1)$. When x_{n-1}, which equals 1, is also considered, the required addition is not done. This is equivalent to subtracting $A \cdot 2^{n-1}$, which is the necessary correction term.

Example 6.2

The following sequential multiplication illustrates case (2) in Table 6.2:

A		1	0	1	1		-5	
X	$\times$	1	1	0	1		-3	
Y		0	$\bar{1}$	1	$\bar{1}$		recoded multiplier	
Add $-A$		0	1	0	1			
Shift		0	0	1	0	1		
Add A	$+$	1	0	1	1			
		1	1	0	1	1		
Shift		1	1	1	0	1	1	
Add $-A$	$+$	0	1	0	1			
		0	0	1	1	1	1	
Shift		0	0	0	1	1	1	1

This multiplication starts from the least significant bit of the multiplier. If started from the most significant bit, a longer adder/subtractor would be needed to allow for carry propagation. Also, note that there is no need to generate the recoded SD multiplier that would require two bits per digit if generated. Instead, the bits of the original multiplier can be scanned, and appropriate control signals for the adder/subtractor can be generated. $\square$

Booth's algorithm can handle two's complement multipliers properly, and consequently if unsigned numbers are to be multiplied, we must add a zero to the left of the multiplier (i.e., $x_n = 0$) to ensure the correctness of the result.

There are two drawbacks to Booth's algorithm. The first is that the number of add/subtract operations is variable, and so is the number of shift operations between two consecutive add/subtract operations. These are very inconvenient when designing a synchronous multiplier. Second, this algorithm becomes inefficient when there are isolated 1's; for example, 001010101(0) is recoded as $01\bar{1}1\bar{1}1\bar{1}1\bar{1}$, requiring eight, instead of four, operations.

The situation can be improved by examining three bits of X at a time rather than two [11]. The bits x_i and x_{i-1} are recoded into y_i and y_{i-1}, while x_{i-2} serves as a reference bit. In a separate step, x_{i-2} and x_{i-3} are recoded into y_{i-2} and y_{i-3}, with x_{i-4} serving as a reference bit. Thus, the groups of three bits each overlap, with the rightmost being $x_1 x_0 (x_{-1})$, the next one being $x_3 x_2 (x_1)$, and so on as shown below:

$$\bullet \quad \bullet \quad \bullet \quad x_7 \quad x_6 \quad x_5 \quad x_4 \quad x_3 \quad x_2 \quad x_1 \quad x_0 \quad (x_{-1})$$

$$y_7 y_6 \qquad y_5 y_4 \qquad y_3 y_2 \qquad y_1 y_0$$

The rules for this radix-4 modified Booth's algorithm are shown in Table 6.3 for all odd values of i, namely, $i = 1, 3, 5, \cdots$. Comparing this algorithm

x_i	x_{i-1}	x_{i-2}	y_i	y_{i-1}	operation	comments
0	0	0	0	0	$+0$	string of zeros
0	1	0	0	1	$+A$	a single 1
1	0	0	$\bar{1}$	0	$-2A$	beginning of 1's
1	1	0	0	$\bar{1}$	$-A$	beginning of 1's
0	0	1	0	1	$+A$	end of 1's
0	1	1	1	0	$+2A$	end of 1's
1	0	1	0	$\bar{1}$	$-A$	a single 0
1	1	1	0	0	$+0$	string of 1's

TABLE 6.3 A radix-4 modified Booth's algorithm.

to the original Booth's algorithm in Table 6.1, we see that an isolated 1 or 0 is handled more efficiently there. If x_{i-1} is an isolated 1, $y_{i-1} = 1$, so only a single operation is needed. A similar simplification occurs if x_{i-1} is an isolated 0 in a string of 1's. In this case, $\cdots 10(1) \cdots$ is recoded into $\cdots \bar{1}1 \cdots$, or, more conveniently, into $\cdots 0\bar{1} \cdots$, and again only a single operation is performed. A simple way to find the required operation is to calculate

$$x_{i-1} + x_{i-2} - 2x_i$$

for odd values of i and represent the result as a 2-bit binary number $y_i y_{i-1}$ in SD notation. The verification of this statement is left as an exercise for the reader.

For $01|01|01|01|(0)$ we now obtain $01|01|01|01|$, and the number of operations remains four, which is the minimum. However, for $00|10|10|10|(0)$ we get $01|0\bar{1}|0\bar{1}|\bar{1}0|$, requiring four, instead of three, operations. Still, compared to the radix-2 Booth's algorithm in Table 6.1, the number of patterns for which the number of partial products is increased, rather than decreased, is smaller. Also, the increase in the number of operations is smaller. In any case, we may design an n-bit synchronous multiplier that generates exactly $n/2$ partial products. Here, too, two's complement multipliers are handled correctly, but we have to make sure that n is even. Otherwise, an extension of the sign bit is required. Also, we need to add a zero to the left of the multiplier if unsigned numbers are multiplied and n is odd. Two zeros must be added if n is an even number.

Example 6.3

A			01	00	01			17	
X	$\times$		11	01	11			-9	
Y			$0\bar{1}$	10	$0\bar{1}$			recoded multiplier	
			$-A$	$+2A$	$-A$			operation	
Add $-A$	$+$		10	11	11				
2-bit Shift		1	11	10	11	11			
Add $2A$	$+$	0	10	00	10				
			01	11	01	11			
2-bit Shift			00	01	11	01	11		
Add $-A$	$+$		10	11	11				
			11	01	10	01	11	-153	

There are $n/2 = 3$ steps in this multiplication and in each step two multiplier bits are dealt with. As a result, all shift operations are two bit position shifts. Also, note that an additional bit for storing the correct sign is required to properly handle the addition of $2A$. □

It is possible to extend the above recoding to three bits at a time, and have overlapping groups of four bits each. The resulting algorithm is called the radix-8 modified Booth's algorithm. In this algorithm, only $n/3$ partial products are generated, but the multiple $3A$ is needed, adding complexity to the basic step. For example, recoding $010(1)$ yields $y_i y_{i-1} y_{i-2} = 011$. A technique for simplifying the generation and accumulation of the multiples $\pm 3A$ has been presented in [7].

An interesting question now arises: what is the minimal number of add/-subtract operations required for a given multiplier? To answer this, we have to find the minimal SD representation of the multiplier; i.e., the one with the smallest number of nonzero digits, $min \sum_{i=0}^{n-1} |y_i|$. It has been shown [20] that a sequence $y_{n-1} y_{n-2} \ldots y_0$ is a minimal representation of an SD number if $y_i \cdot y_{i-1} = 0$ for $1 \leq i \leq n-1$, given that the most significant bits can satisfy $y_{n-1} \cdot y_{n-2} \neq 1$. To see the reason behind this condition consider, for example, the representation of 7 with only three bits; here 111 is a minimal representation although $y_i \cdot y_{i-1} \neq 0$. In practice, for any multiplier X, we can always add a 0 to its left to make sure that the above condition is satisfied.

The algorithm for obtaining the minimal representation of X is described next. The multiplier bits are examined from right to left, one bit at a time with the next bit to the left (i.e., x_{i+1}) serving as a reference bit. To correctly handle a single 0 within a string of 1's (and similarly, a single 1 within a string of 0's) we need information on the kind of string that exists to the right of the current position. For this purpose we use a "carry" bit (0 for 0's and 1 for 1's). This algorithm is called *canonical recoding* and its rules are summarized in Table 6.4, where c_i is the previous "carry" and c_{i+1} is the next "carry."

As before, the recoded multiplier (after canonical recoding) can be used without any correction steps if the original multiplier is represented in two's complement. Here, we have to extend the sign bit x_{n-1}, obtaining $x_{n-1} x_{n-1} x_{n-2} \cdots x_0$. Canonical recoding can also be expanded to generate two or more bits at a time. The multiples of A needed in the case of two bits are $\pm A$ and $\pm 2A$.

x_{i+1}	x_i	c_i	y_i	c_{i+1}	Comments
0	0	0	0	0	string of 0's
0	1	0	1	0	a single 1
1	0	0	0	0	string of 0's
1	1	0	$\bar{1}$	1	beginning of 1's
0	0	1	1	0	end of 1's
0	1	1	0	1	string of 1's
1	0	1	$\bar{1}$	1	a single 0
1	1	1	0	1	string of 1's

TABLE 6.4 Canonical recoding.

x_{i+1}	x_i	x_{i-1}	Operation	Comments
0	0	0	$+0$	string of 0's
0	0	1	$+2A$	end of 1's
0	1	0	$+2A$	a single 1
0	1	1	$+4A$	end of 1's
1	0	0	$-4A$	beginning of 1's
1	0	1	$-2A$	a single 0
1	1	0	$-2A$	beginning of 1's
1	1	1	$+0$	string of 1's

TABLE 6.5 An alternate 2-bit-at-a-time multiplication algorithm.

The main disadvantage of canonical recoding is that the bits of the multiplier are generated *sequentially*, while in the original and modified Booth's algorithms we may generate the bits simultaneously (there is no "carry" propagation). This implies that in the latter case, we can generate all partial products in parallel, and then use a fast multi-operand adder.

Another drawback to canonical recoding is that, like Booth's algorithm, in order to take full advantage of the minimum number of add/subtract operations the number of these operations must be variable, as must be the length of the shift operations. This is difficult to implement, and we would prefer to have uniform shifts. This implies that the number of partial products will always be $n/2$, although canonical recoding can lead to a much smaller number of operations.

The radix-4 modified Booth's algorithm in Table 6.3 is not the only way of reducing the number of partial products, while still having uniform shifts of two bits each. Instead of using the next bit to the right (x_{i-2}) as a reference bit when examining $x_i x_{i-1}$, we can use the next bit to the left (x_{i+1}). The rules for this multiplication algorithm are summarized in Table 6.5 where, as before, i is an odd number. The multiples of A that are needed are $\pm 2A$ and $\pm 4A$, and they can be easily generated using shifts. The multiple $4A$ must be generated when $(x_{i+1})x_i x_{i-1} = (0)11$ to take care of the end of group of 1's. This can not be done at the time when the bits $(x_{i+3})x_{i+2}x_{i+1}$ are examined, since they have a zero in the rightmost position. As a result, this algorithm is not a recoding of the multiplier, as we cannot express 4 in two bits. The number of partial products is always $n/2$. As for canonical recoding, two's complement multipliers can be handled by extending the sign bit. Also, if unsigned numbers are multiplied, one or two zeros must be added to the left of the multiplier.

Example 6.4

For the multiplier 01101110, the following partial products are generated:

$$(0) \quad 01 \quad\quad 10 \quad\quad 11 \quad\quad 10$$
$$+2A \quad -2A \quad +4A \quad -2A$$

This translates to the SD number $010\bar{1}100\bar{1}0$, which is not a minimal representation, since it includes two adjacent nonzero digits. Employing the canonical recoding summarized in Table 6.4 yields $0100\bar{1}00\bar{1}0$, which is a minimal representation.   □

For the rightmost pair x_1x_0 , if $x_0 = 1$ it is considered a continuation of a string of 1's that never really started, and therefore no subtraction took place. For example, the multiplier 01110111 results in the following partial products:

$$
\begin{array}{cccc}
01 & 11 & 01 & 11 \\
+2A & +0 & -2A & +0 \\
\end{array}
$$

instead of
$$
\begin{array}{cccc}
+2A & +0 & -2A & -A \\
\end{array}
$$

This can be corrected by setting the initial partial product to be $-A$ instead of 0 whenever $x_0 = 1$. All four possible cases are listed in Table 6.6.

x_2	x_1	x_0	Operation
0	0	1	$+2A - A = A$
0	1	1	$+4A - A = 3A$
1	0	1	$-2A - A = -3A$
1	1	1	$0 - A = -A$

TABLE 6.6   The handling of x_1x_0 with $x_0 = 1$ in the algorithm of Table 6.5.

Example 6.5

We repeat the previous example (where the radix-4 modified Booth's algorithm was used) and obtain

$$
\begin{array}{llcccccc}
A   &   &   & 01 & 00 & 01 &   &   & 17 \\
X   & \times & (1) & 11 & 01 & 11 &   &   & -9 \\
           &   &   & 0 & -2A & 0 &   &   & \text{Operation} \\
\hline
\text{Initial} -A & &   & 10 & 11 & 11 &   &   & \\
\text{Add } 0   & + &   & 00 & 00 & 00 &   &   & \\
\hline
           &   &   & 10 & 11 & 11 &   &   & \\
\text{2-bit Shift} & & 1 & 11 & 10 & 11 & 11 &   & \\
\text{Add} -2A   & + & 1 & 01 & 11 & 10 &   &   & \\
\hline
           &   & 1 & 01 & 10 & 01 & 11 &   & \\
\text{2-bit Shift} & & & 11 & 01 & 10 & 01 & 11 & \\
\text{Add } 0   & + & & 00 & 00 & 00 &   &   & \\
\hline
           &   &   & 11 & 01 & 10 & 01 & 11 & -153 \\
\end{array}
$$

Note that the multiplier's sign bit had to be extended in order to decide that no operation is needed for the first pair of multiplier bits. Also, as in the previous example, an additional bit for holding the correct sign is needed, because of multiples like $-2A$.   □

The method summarized in Table 6.5 can also be extended to three bits or more at each step. However, as in the radix-8 modified Booth's algorithm, multiples of A like $3A$ or even $6A$ are needed, and unless those are prepared in advance and stored somewhere, we have to perform two additions in a single step. For example, for $(0)101$ we need $8 - 2 = 6$, and for $(1)001$, $-8 + 2 = -6$.

6.2 IMPLEMENTING LARGE MULTIPLIERS USING SMALLER ONES

If an $n \times n$ bit multiplier is implemented as a single integrated circuit, we can use several such circuits for implementing larger multipliers. A $2n \times 2n$ bit multiplier can be constructed out of four $n \times n$ bit multipliers. This is based on the following equation:

$$A \cdot X = (A_H \cdot 2^n + A_L) \cdot (X_H \cdot 2^n + X_L) = A_H \cdot X_H \cdot 2^{2n} + (A_H \cdot X_L + A_L \cdot X_H) \cdot 2^n + A_L \cdot X_L$$
(6.2)

where A_H and A_L are the most and least significant halves of A, respectively, and X_H and X_L are likewise for X.

The four partial products of length $2n$ bits each should be correctly aligned before being added, as shown in Figure 6.1(a). A more convenient arrangement is shown in Figure 6.1(b). This last arrangement gives the minimum height of

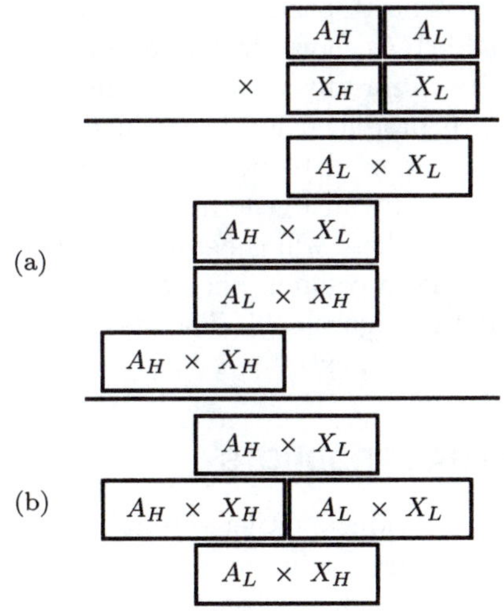

FIGURE 6.1   Aligning the four partial products in Equation (6.2).

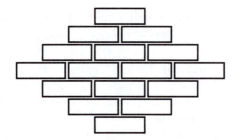

FIGURE 6.2   Aligning the 16 partial products.

the matrix of numbers to be added requiring one level of carry-save addition and a CPA (carry-propagating adder). Note that the n least significant bits are already bits of the final product, and no further addition is needed. The $2n$ bits in the center have to be added by a $2n$-bit CSA, whose outputs are connected to a CPA. The n most significant bits have to be connected to the same CPA, since the center bits may generate a carry into the most significant bits. Thus, a $3n$-bit CPA is needed.

The idea of decomposing a large multiplier into smaller ones can be further extended. First, the basic multiplier used as a building block can be an $n \times m$ bit multiplier, with $n \neq m$. Second, multipliers larger than $2n \times 2m$ can be implemented. For example, a $4n \times 4n$ bit multiplier can be implemented using available $n \times n$ bit multipliers. A $4n \times 4n$ bit multiplier requires four $2n \times 2n$ bit multipliers, which in turn require four $n \times n$ bit multipliers each, for a total of 16 $n \times n$ bit multipliers. The 16 partial products generated this way have to be aligned before being added, as shown in Figure 6.2. Similar arrangements of partial products can be drawn for any $kn \times kn$ bit multiplier with an integer k.

After aligning the 16 products, as shown in Figure 6.2, we have up to seven bits in one column that need to be added. To add seven operands we may use a set of (7,3) counters, which generate three operands, to be added by another set of (3,2) counters. These will generate two operands, to be added by a CPA. Another possibility is to combine the two sets of counters into a set of (7;2) compressors, depicted in Figure 5.26. The task of selecting an economical multi-operand adder is discussed next.

6.3 ACCUMULATING THE PARTIAL PRODUCTS

After generating the partial products either through one of the algorithms discussed in Section 6.1 or by using smaller multipliers, as in Section 6.2, we must accumulate all these partial products to obtain the final product. If a high-speed accumulation of partial products is desired, a fast multi-operand adder should be employed. Such multi-operand adders, using different types of parallel counters,

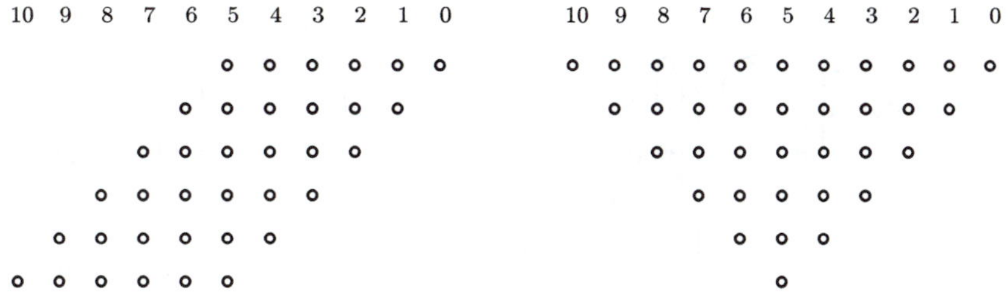

(a) Original matrix of 36 bits.   (b) Reorganized matrix of bits.

FIGURE 6.3   Six partial products to be added.

have been described in Chapter 5. We should, however, take advantage of the particular form of the partial products to be added and reduce the hardware complexity of the multi-operand adder. The partial products to be added have a smaller number of bits than the final product, and they have to be aligned before being added. Thus, we can expect to see many columns that include fewer bits than the total number of partial products, requiring simpler counters for their addition.

Consider, for example, the six partial products that are generated when multiplying two unsigned operands of length 6 bits each, using the simple one-bit-at-a-time algorithm. The matrix of partial product bits to be added is shown in Figure 6.3(a). These six operands can be added using the three-level carry-save tree shown in Figure 5.24. The number of (3,2) counters can, however, be substantially reduced by taking advantage of the fact that all columns but one in Figure 6.3(a) contain fewer than six bits. To simplify the task of deciding how many counters are needed we can redraw the matrix of bits to be added, as depicted in Figure 6.3(b).

To further reduce the hardware complexity we also allow the use of half adders (HAs) in addition to full adders (FAs). An HA, which can be called a (2,2) counter, has a lower hardware complexity than an FA. Figure 6.4 depicts the (3,2) and (2,2) counters that can be used in order to reduce the number of operands from 6 to 2. These two operands are then added through a CPA. In this figure, a vertical block containing three bits represents a (3,2) counter, while a vertical block containing two bits represents a (2,2) counter. The horizontal blocks in Figure 6.4(b) show the outputs of the (3,2) and (2,2) counters in Figure 6.4(a). For example, the horizontal block in columns 2 and 3 contains the two outputs of the (3,2) counter in column 2 of Figure 6.4(a). The number of levels in the carry-save addition is still 3, but the number of counters is substantially smaller than that needed in the general case (see Figure 5.24).

The number of counters can be further reduced by employing the idea mentioned in Chapter 5 of reducing the number of bits in each column to the

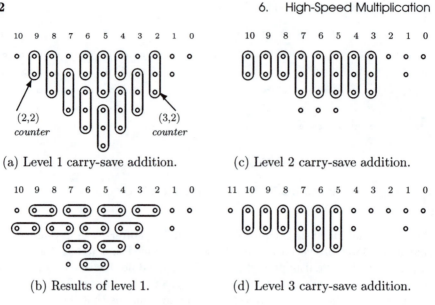

(a) Level 1 carry-save addition.

(c) Level 2 carry-save addition.

(b) Results of level 1.

(d) Level 3 carry-save addition.

FIGURE 6.4   Reduction of the six partial products.

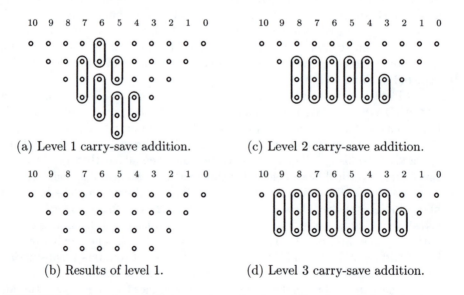

(a) Level 1 carry-save addition.

(c) Level 2 carry-save addition.

(b) Results of level 1.

(d) Level 3 carry-save addition.

FIGURE 6.5   An alternate scheme for reduction of the six partial products.

closest element of the series $3, 4, 6, 9, 13, 19, \cdots$ [4]. This is shown in Figure 6.5 where for example, in Figure 6.5(a), column 5, the smallest number of counters which will reduce the number of bits to four, is used. Overall, the scheme in Figure 6.5 requires fifteen (3,2) counters and five (2,2) counters, compared to the sixteen (3,2) counters and nine (2,2) counters needed in Figure 6.4. The savings are even more substantial when larger multipliers are designed.

The above discussion is restricted to unsigned numbers. If some of the partial products are negative numbers represented in two's complement, we need to modify the matrix of bits shown in Figure 6.3(a). Specifically, all sign bits must be properly extended before the addition of the partial products takes place, yielding the matrix shown in Figure 6.6. The number of bits in row 1 is now 11 instead of 6, and so on. This extension significantly increases the hardware complexity of the multi-operand adder required. If two's complement numbers are obtained by generating the one's complement and then adding a carry to the least significant bit, the matrix will have to be increased even further.

We may minimize the increase in complexity by realizing that the two's complement number

$$s \; s \; s \; s \; s \; s \; z_4 \; z_3 \; z_2 \; z_1 \; z_0$$

whose value is

$$-s \cdot 2^{10} + s \cdot 2^9 + s \cdot 2^8 + s \cdot 2^7 + s \cdot 2^6 + s \cdot 2^5 + z_4 \cdot 2^4 + z_3 \cdot 2^3 + z_2 \cdot 2^2 + z_1 \cdot 2^1 + z_0$$

can be replaced by

$$0 \; 0 \; 0 \; 0 \; 0 \; (-s) \; z_4 \; z_3 \; z_2 \; z_1 \; z_0$$

since

$$-s \cdot 2^{10} + s \cdot (2^9 + 2^8 + 2^7 + 2^6 + 2^5) = -s \cdot 2^{10} + s \cdot (2^{10} - 2^5) = \; -s \cdot 2^5.$$

To represent the value $-s$ in column 5 (in Figure 6.6), we complement the original sign digit s to obtain $(1 - s)$, and add 1. We get the $-s$ as required

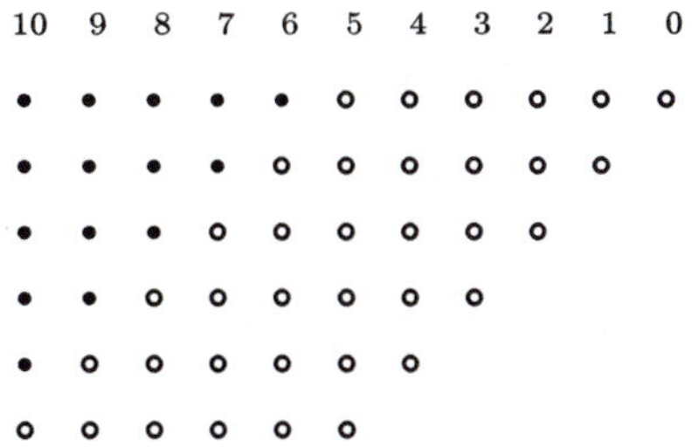

FIGURE 6.6   Six signed partial products to be added (full circles indicate extended sign bits).

```
10   9   8   7   6   5   4   3   2   1   0
                     1
                    S̄₁  o   o   o   o   o
                 S̄₂  o   o   o   o   o
              S̄₃  o   o   o   o   o
           S̄₄  o   o   o   o   o
        S̄₅  o   o   o   o   o
     S̄₆  o   o   o   o   o
```

FIGURE 6.7   The modified array of six signed partial products.

along with a carry of 1 into column 6. The latter will serve as the extra 1 needed in column 6 to deal with the sign bit of the second partial product. Another carry-out will be generated in column 6 and so on. The resulting matrix of bits is shown in Figure 6.7. This new matrix has fewer bits than that in Figure 6.6 but has a higher maximum height (7 instead of 6, in column 5). We can eliminate the extra 1 in column 5 if we place the two sign bits s_1 and s_2 in the same column, since

$$(1 - s_1) + (1 - s_2) = 2 - s_1 - s_2.$$

The 2 is carried out to the next column, leaving behind $-s_1$ and $-s_2$. An extra 1 in this column is no longer required. Placing the two sign bits in the same column can be achieved by first extending the sign bit s_1 in one position, as shown in Figure 6.8. The maximum column height is now back to 6.

If the negative partial products are obtained by first generating the one's complement and then adding a carry to the least significant bit, these extra carries can then be added to the matrix, as shown in Figure 6.9. The full circles indicate that the complements of the corresponding bits are taken whenever $s_i = 1$. The extra s_6 in column 5 increases the maximum column height to 7. However, if the last partial product is always positive (i.e., the multiplier is always positive), this s_6 can be eliminated.

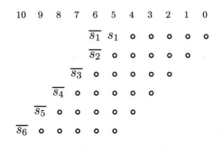

```
10   9   8   7   6   5   4   3   2   1   0
                 S̄₁  s₁  o   o   o   o   o
                 S̄₂  o   o   o   o   o
              S̄₃  o   o   o   o   o
           S̄₄  o   o   o   o   o
        S̄₅  o   o   o   o   o
     S̄₆  o   o   o   o   o
```

FIGURE 6.8   Further modified array of six signed partial products.

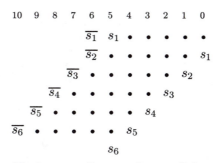

FIGURE 6.9   The modified array with negative partial products represented in one's complement.

Example 6.6

In this example, the negative partial products are generated as a result of a recoded multiplier using canonical recoding. If all sign bits are extended, the following matrix is obtained, where all replicated sign bits are shown in bold face:

A						0	1	0	1	1	0	22	
X			$\times$			0	0	1	0	1	1	11	
Y						0	1	0	$\bar{1}$	0	$\bar{1}$	Recoded multiplier	
1	**1**	**1**	**1**	**1**	**1**	0	1	0	1	0		
0	**0**	**0**	**0**	**0**	**0**	0	0	0	0			
1	**1**	**1**	**1**	0	1	0	1	0				
0	**0**	**0**	**0**	0	0	0	0					
0	**0**	1	0	1	1	0						
0	0	0	0	0	0							
0	0	0	1	1	1	1	0	0	1	0		

A smaller matrix of bits to be added is obtained if we follow the scheme illustrated in Figure 6.8:

10	9	8	7	6	5	4	3	2	1	0
				0	1	0	1	0	1	0
				1	0	0	0	0	0	
			0	0	1	0	1	0		
		1	0	0	0	0	0			
	1	1	0	1	1	0				
1	0	0	0	0	0					
0	0	0	1	1	1	1	0	0	1	0

If the negative partial are generated using one's complement and a carry into the least significant position, then the resulting matrix becomes

10	9	8	7	6	5	4	3	2	1	0
				0	1	0	1	0	0	1
				1	0	0	0	0	0	1
		0	0	1	0	0	1	**0**		
	1	0	0	0	0	0	1			
	1	1	0	1	1	0	**0**			
1	0	0	0	0	0	**0**				
					0					
0	0	0	1	1	1	1	0	0	1	0

$\square$

If we use the modified radix-4 Booth algorithm for generating the partial products, the resulting matrices, corresponding to Figures 6.7, 6.8 and 6.9, are shown in Figure 6.10(a), (b) and (c), respectively. Note that in Figure 6.10(a), the carry generated by adding 1 to $\overline{s_1} = (1 - s_1)$ for the first partial product is not positioned in the same column as that of $\overline{s_2} = (1 - s_2)$, the complement of the sign bit of the second partial product. We need, therefore, to put an extra 1 in column 7 which, together with the carry generated by column 6, produces the necessary 1 in column 8.

Example 6.7

We repeat the multiplication from the previous example but now use the radix-4 modified Booth's algorithm resulting in three partial products. The recoded multiplier turns out in this case to be exactly the same; i.e., $010\overline{1}0\overline{1}$. The second scheme (see Figure 6.10(b)) results in

9	8	7	6	5	4	3	2	1	0
		0	**1**	**1**	0	**1**	0	**1**	0
	1	**0**	0	1	0	1	0		
1	1	0	1	1	0				
0	0	1	1	1	1	0	0	1	0

If the third scheme (see Figure 6.10(c)) is followed, the resulting matrix is

9	8	7	6	5	4	3	2	1	0
		0	**1**	**1**	0	**1**	0	0	**1**
	1	**0**	0	1	0	0	**1**		**1**
1	1	0	1	1	0		**1**		
					0				
0	0	1	1	1	1	0	0	1	0

$\square$

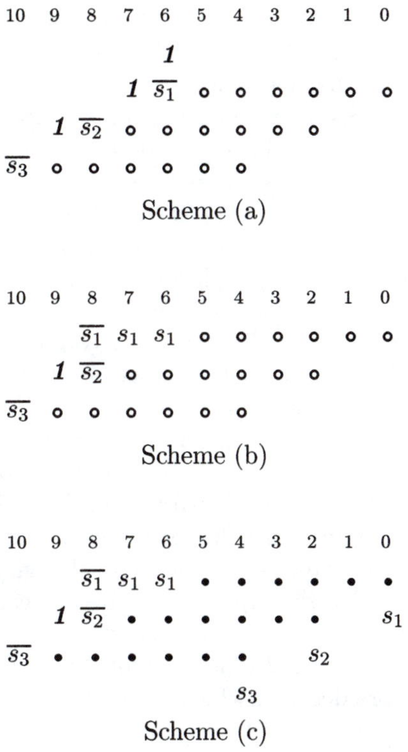

```
10  9  8  7  6  5  4  3  2  1  0
                1
             1  s̄₁  o  o  o  o  o  o  o
          1  s̄₂  o  o  o  o  o  o  o
       s̄₃  o  o  o  o  o  o  o
```

Scheme (a)

```
10  9  8  7  6  5  4  3  2  1  0
          s̄₁ s₁ s₁  o  o  o  o  o  o
       1  s̄₂  o  o  o  o  o  o  o
    s̄₃  o  o  o  o  o  o  o
```

Scheme (b)

```
10  9  8  7  6  5  4  3  2  1  0
          s̄₁ s₁ s₁  •  •  •  •  •  •
       1  s̄₂  •  •  •  •  •  •          s₁
    s̄₃  •  •  •  •  •  •          s₂
                            s₃
```

Scheme (c)

FIGURE 6.10 Three schemes for an array of three (radix-4 modified Booth algorithm) partial products.

6.4 ALTERNATIVE TECHNIQUES FOR PARTIAL PRODUCT ACCUMULATION

Several modifications to the basic tree structure for partial product accumulation have been suggested and implemented. The purpose of these techniques is to reduce the number of levels in the tree (and, as a result, speed up the accumulation) and/or achieve a more regular design. Tree structures usually have very irregular interconnects. This irregularity complicates the implementation and, more importantly, irregular structures result in area-inefficient layouts, especially when a rectangular-shaped layout is sought. Notice also that a smaller number of levels results in less irregularity.

The number of levels in the tree can be lowered by using a reduction rate higher than 3:2. A reduction rate of 2:1 can be achieved if the carry-save adders are replaced by adders for binary SD numbers described in Section 2.4. Like the carry-save adder, the SD adder generates the sum of its two operands in constant time (independent of the number of bits), since the carry is allowed to propagate at most one position. The number of levels in the SD adder tree is smaller and, in addition, the tree produces a single result rather than the two results of the ordinary CSA tree (see for example, Figure 5.24). However, the result of the SD adder tree is still in SD representation and, consequently, in most cases, a conversion to two's complement representation is needed. This conversion is done by forming two sequences. The first sequence, denoted by Z^+, is created by replacing each negative digit of the SD number by zero. The second sequence, denoted by Z^-, replaces each negative digit of the original SD number with its absolute value, and each positive digit by zero. Then, the difference $Z^+ - Z^-$ is found by adding the two's complement of Z^- to Z^+ using a carry-propagating adder. Hence, a final stage of a CPA is needed here, as it is needed in the ordinary CSA tree.

Another advantage of an SD adder tree over a CSA tree is that there is no need for a sign bit extension when negative partial products are to be added. SD numbers simply do not require a separate sign bit. The major disadvantage of the SD adder is that its design is more complex, consuming more gates and consequently a larger chip area, since each signed digit requires two ordinary bits (or a multiple-valued logic implementation that can provide three values per digit, corresponding to $-1, 0$, and 1). As a result, a more careful comparison between the CSA tree and the SD adder tree for the particular given technology must be performed before deciding which to employ.

Example 6.8

A 32×32 multiplier based on the radix-4 modified Booth's algorithm generates 16 partial products to be accumulated and consequently requires a CSA tree with six levels (see Table 5.1), but needs an SD adder tree with only four levels. Some sophisticated logic design techniques and layout schemes can be employed, resulting in less area-consuming implementations [10]. □

The same reduction rate of 2:1 can be achieved without resorting to SD representations by using (4;2) compressors, shown in Figure 6.11. Similarly to the (7;2) compressor in Figure 5.26, the (4;2) compressor must be designed so that c_{out} is not a function of c_{in}, in order to avoid a ripple-carry effect. Also, the (4;2) compressor may be implemented as a multi-level circuit with a smaller overall delay compared to the implementation based on two (3,2) counters, as in Figure 6.11. One such implementation is shown in Figure 6.12 with a delay of three exclusive-or gates between the inputs (x_1, x_2, x_3 and x_4) and the output S.

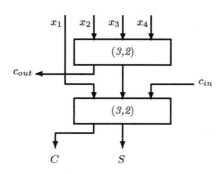

FIGURE 6.11   A (4;2) compressor.

Implementing the (4;2) compressor with two (3,2) counters, as shown in Figure 6.11, will result in a delay of four exclusive-or gates. Thus, the delay of the implementation in Figure 6.12 is expected to be 25% lower than that of the implementation in Figure 6.11.

Other multi-level implementations of a (4;2) compressor are possible. All such implementations must satisfy the following arithmetic equation:

$$x_1 + x_2 + x_3 + x_4 + c_{in} = S + 2(C + c_{out}),$$

and c_{out} should not depend on c_{in}, to avoid horizontal rippling of carries. The truth table for such implementations is summarized in Table 6.7, where a, b, c, d, e and f are Boolean variables. The implementation in Figure 6.12 corresponds to the setting $a = b = c = 1$ and $d = e = f = 0$.

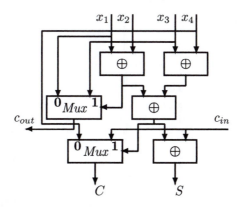

FIGURE 6.12   An implementation of a (4;2) compressor.

x_1	x_2	x_3	x_4	c_{out}	C	S	x_1	x_2	x_3	x_4	c_{out}	C	S
0	0	0	0	0	0	c_{in}	1	0	0	0	0	c_{in}	$\overline{c_{in}}$
0	0	0	1	0	c_{in}	$\overline{c_{in}}$	1	0	0	1	d	$\bar{d}$	c_{in}
0	0	1	0	0	c_{in}	$\overline{c_{in}}$	1	0	1	0	e	$\bar{e}$	c_{in}
0	0	1	1	a	$\bar{a}$	c_{in}	1	0	1	1	1	c_{in}	$\overline{c_{in}}$
0	1	0	0	0	c_{in}	$\overline{c_{in}}$	1	1	0	0	f	$\bar{f}$	c_{in}
0	1	0	1	b	$\bar{b}$	c_{in}	1	1	0	1	1	c_{in}	$\overline{c_{in}}$
0	1	1	0	c	$\bar{c}$	c_{in}	1	1	1	0	1	c_{in}	$\overline{c_{in}}$
0	1	1	1	1	c_{in}	$\overline{c_{in}}$	1	1	1	1	1	1	c_{in}

TABLE 6.7   The truth table for a (4;2) compressor.

An adder tree that uses (4;2) compressors will have a more regular structure and may have a lower delay than an ordinary CSA tree made of (3,2) counters. Table 6.8 compares the delays of carry-save trees using either (3,2) counters or (4;2) compressors. Since the delay of a (4;2) compressor is 1.5 times that of a (3,2) counter, the number of levels of (4;2) compressors in column 3 is multiplied by 1.5 to yield the equivalent delay in column 4. Note that the equivalent delay of a carry save tree using (4;2) compressors (column 4) is not always smaller than that of a carry save tree using (3,2) counters (column 2). For example, for nine partial products, (3,2) counters will yield a carry save tree with an overall lower delay. Various other counters and compressors can be employed in the implementation of the addition tree for the partial product accumulation; for example, (7,3) counters [13].

Number of operands	Number of levels using (3,2)	Number of levels using (4;2)	Equivalent delay
3	1	1	1.5
4	2	1	1.5
5 – 6	3	2	3
7 – 8	4	2	3
9	4	3	4.5
10 – 13	5	3	4.5
14 – 16	6	3	4.5
17 – 19	6	4	6
20 – 28	7	4	6
29 – 32	8	4	6
33 – 42	8	5	7.5

TABLE 6.8   Comparing the delays of carry-save adder using either (3,2) counters or (4;2) compressors.

Several other techniques have been suggested to modify the structure of CSA trees which use (3,2) counters, in order to achieve a more regular and less area-consuming layout. Such modified tree structures may require a somewhat larger number of CSA levels with a larger overall delay. Two such techniques are described next. The first one defines balanced delay trees [24] (see also [19]) while the second one defines overturned-stairs trees [15]. Figure 6.13 illustrates the structure of the bit-slices for these two techniques and compares them to the corresponding Wallace tree bit slice. All the bit-slices in Figure 6.13 are for 18 operands (partial products) which may be generated by a high radix multiplication algorithm (e.g., a radix-4 modified Booth's algorithm). In this case, the 18 downward triangles in Figure 6.13 represent multiplexers that select the suitable multiple of the multiplicand. The rectangles represent (3,2) counters, and the numbers on these counters indicate the delay experienced by the input operands. Thus, after $6\Delta_{FA}$, two results are produced by the Wallace and the overturned-stairs trees; the balanced tree requires $7\Delta_{FA}$.

Note that all three tree structures have fifteen outgoing carries and fifteen incoming carries, and each outgoing carry is aligned with its corresponding incoming carry (from the previous bit slice), so that adjacent bit-slices abut. The incoming carries are routed to different (3,2) counters so that all the inputs to a counter are valid before or at the necessary time. Only for the balanced tree are all fifteen incoming carries generated exactly when they are required, since all paths are balanced. In the other two trees, there are counters for which not all incoming carries are generated simultaneously. For example, the bottom counter in the overturned-stairs tree has incoming carries whose associated delays are $4\Delta_{FA}$ and $5\Delta_{FA}$.

The three tree structures also differ in the number of required wiring tracks between adjacent bit-slices; these in turn, affect the layout area. The Wallace tree requires six wiring tracks; the overturned-stairs and the balanced tree require three and two tracks, respectively. Note the inherent tradeoff between size and speed. A Wallace tree guarantees the lowest overall delay but requires the highest number of wiring tracks (on the order of $\log N$, where N is the number of inputs). The balanced tree, on the other hand, requires the smallest number of wiring tracks but has the highest overall delay.

The balanced and overturned-stairs trees have a regular structure and can be designed in a systematic way. This is difficult to see from Figure 6.13, but it can be concluded from Figure 6.14, which shows the complete structure of the two trees as well as that of the corresponding Wallace tree. The building blocks of the balanced and overturned-stairs trees are indicated with dotted lines in Figure 6.14. The exact details of the recursive construction of the two types of regular trees, and some variations of them, can be found in [24] and [15].

When determining the final layout of a CSA tree, care must be taken to make sure that wires connecting the inputs to a carry-save adder have roughly

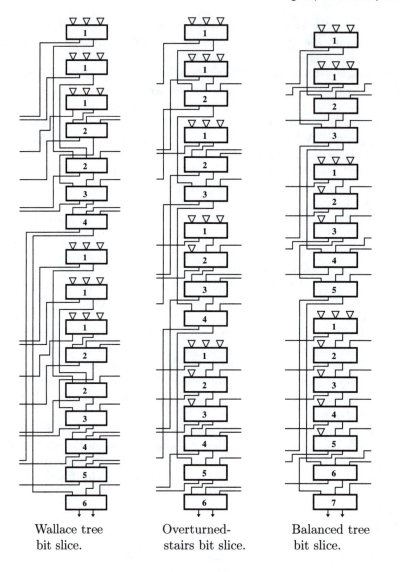

Wallace tree
bit slice.

Overturned-
stairs bit slice.

Balanced tree
bit slice.

FIGURE 6.13   Three CSA tree bit-slices for 18 operands (downward triangles are multiplexers).

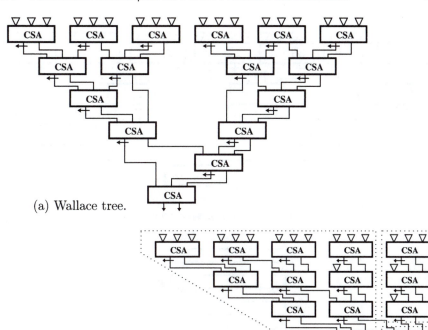

(a) Wallace tree.

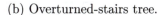

(b) Overturned-stairs tree.

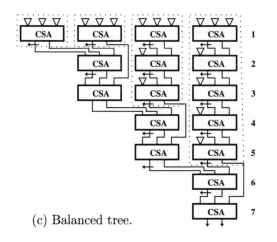

(c) Balanced tree.

FIGURE 6.14   Wallace, overturned-stairs and balanced trees for 18 operands (downward triangles are multiplexers).

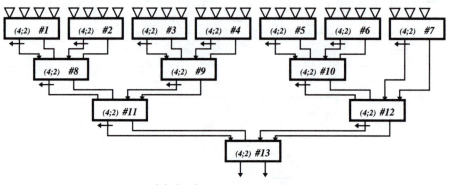

(a) (4;2) compressor tree.

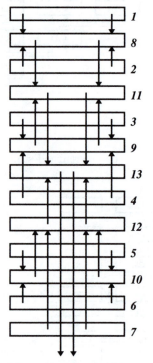

(b) Simplified layout of the (4;2) compressor tree.

FIGURE 6.15   A (4;2) compressor tree for 27 partial products and its layout.

the same length otherwise the delay balanced paths will no longer be balanced. Consider, for example, a CSA tree for 27 operands (27 partial products obtained from a 53-bit multiplier using the radix-4 modified Booth algorithm). A CSA tree constructed out of (4;2) compressors is shown in Figure 6.15(a), and the corresponding layout is shown in Figure 6.15(b) [25]. Note that the bottom compressor (#13) is located in the middle so that compressors #11 and #12 are roughly at the same distance from it. Compressor #11 in turn has equal length wires from #8 and #9 and so on.

6.5 FUSED MULTIPLY-ADD UNIT

A fused multiply-add unit performs the multiplication $A \times B$ followed immediately by an addition of the product and a third operand C so that the calculation of $A \times B + C$ is done as a single and indivisible operation. Clearly, such a unit is capable of performing multiply only, by setting $C = 0$, and add (or subtract) only by setting, for example, $B = 1$.

A fused multiply-add unit can reduce the overall execution time of chained multiply and then add/subtract operations. An example of a case when such chained multiply and add are useful is in the evaluation of a polynomial $a_n x^n + a_{n-1} x^{n-1} + \cdots + a_0$ through $[(a_n x + a_{n-1})x + a_{n-2}]x + \cdots$. On the other hand, independent multiply and add operations can not be performed in parallel.

Another advantage of a fused multiply-add unit, compared to separate multiplier and adder, arises when executing floating-point operations since rounding is performed only once for the result $A \times B + C$ rather then twice (for the multiply and then for the add). Since rounding may introduce computation errors, reducing the number of roundings may have a positive effect on the overall error. In the design reported in [14], this additional accuracy was helpful when producing a correctly rounded quotient in the divide by reciprocation algorithm (see Section 8.2).

Figure 6.16 shows an implementation of a fused multiply-add unit for floating-point computations. Here, A, B and C are the significands while E_A, E_B and E_C are the exponents of the operands, respectively. The CSA tree generates all the partial products and performs their carry-save accumulation to produce two results which are then added with the properly aligned operand C. The adder accepts three operands and therefore, must first reduce them to two (using (3,2) counters) and then perform carry-propagate addition. The steps of post-normalization and rounding are executed next.

The design illustrated in Figure 6.16 employs two techniques in order to reduce the overall execution time. First, the leading zero anticipator circuit uses the propagate and generate signals produced by the adder (see Section 5.2), to predict the type of shift which will be needed in the post-normalization step.

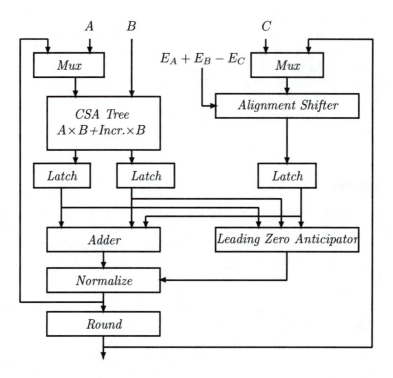

FIGURE 6.16   Fused floating-point multiply-add unit.

This circuit operates in parallel to the addition itself so that the delay of the normalization step is shorter. Second, and more importantly, the alignment of the significand C in $E_A + E_B - E_C$ is done in parallel to the multiplication of A and B. Normally, in a floating-point addition, we align the significand of the smaller operand (i.e., the operand with the smaller exponent). This will imply that if the product $A \times B$ is smaller than C, we will have to shift the product after it has been generated, introducing additional delay. We prefer instead to always align C even if it is larger than $A \times B$, to allow the shift to be performed in parallel to the multiplication. To achieve this, we must allow C to shift either to the right (as is traditionally done) or to the left, the direction dictated by whether the result of $E_A + E_B - E_C$ is either positive or negative, respectively. If we allow C to be shifted to the left we must increase the total number of bits in the adder. For example, if all operands are floating-point numbers in the long IEEE format, the possible range of C relative to the product $A \times B$ is shown as follows:

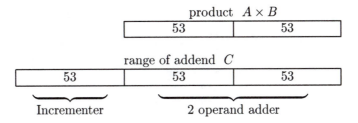

This is the range for $53 \geq E_A + E_B - E_C \geq -53$. If $E_A + E_B - E_C \geq 54$, the bits of C which are shifted further to the right will be replaced by a sticky bit, and if $E_A + E_B - E_C \leq -54$, all the bits of $A \times B$ will be replaced by a sticky bit. The overall penalty is thus a 50% increase in the width of the adder which, in turn, will increase the execution time of the adder. Note however, that the top 53 bits of the adder need only be capable of incrementing the original contents of the 53 bits if a carry propagates from the lower 106 bits.

The path from the output of the rounding circuit in Figure 6.16 to the multiplexer on the right is used when performing a calculation like $(X \times Y + Z) + A \times B$. The path from the output of the normalization circuit to the multiplexer on the left is used when performing a calculation like $(X \times Y + Z) \times B + C$. In this case, the rounding step for $(A \times B + C)$ is performed at the same time as the multiplication by D, by adding the partial product $Incr. \times D$ to the CSA tree.

6.6 ARRAY MULTIPLIERS

The two basic operations, the generation of partial products and their summation, may be merged. In this way, we avoid the overhead that is due to the separate controls of these two operations, and we thus speed up the multiplication.

Such multipliers, which consist of identical cells, each capable of forming a new partial product and adding it to the previously accumulated partial product, are called *iterative array multipliers* or simply *array multipliers*. Clearly, any gain in speed is obtained at the expense of extra hardware. Another important characteristic of array multipliers is that they can be implemented so as to support a high rate of pipelining.

To illustrate the operation of an array multiplier, examine the 5×5 parallelogram shown in Figure 6.17, which contains all 25 partial product bits of the form $a_i \cdot x_j$ properly aligned. A straightforward implementation of an array multiplier adds the first two partial products (i.e., $a_4 \cdot x_0, a_3 \cdot x_0, \cdots a_0 \cdot x_0$ and $a_4 \cdot x_1, a_3 \cdot x_1, \cdots a_0 \cdot x_1$) in row one after proper alignment. The results of the first row are then added to $a_4 \cdot x_2, a_3 \cdot x_2, \cdots a_0 \cdot x_2$ in the second row, and so on. The basic cell for such an array multiplier is an FA accepting one bit of the new

					a_4	a_3	a_2	a_1	a_0
				$\times$	x_4	x_3	x_2	x_1	x_0
					$a_4 \cdot x_0$	$a_3 \cdot x_0$	$a_2 \cdot x_0$	$a_1 \cdot x_0$	$a_0 \cdot x_0$
				$a_4 \cdot x_1$	$a_3 \cdot x_1$	$a_2 \cdot x_1$	$a_1 \cdot x_1$	$a_0 \cdot x_1$	
			$a_4 \cdot x_2$	$a_3 \cdot x_2$	$a_2 \cdot x_2$	$a_1 \cdot x_2$	$a_0 \cdot x_2$		
		$a_4 \cdot x_3$	$a_3 \cdot x_3$	$a_2 \cdot x_3$	$a_1 \cdot x_3$	$a_0 \cdot x_3$			
	$a_4 \cdot x_4$	$a_3 \cdot x_4$	$a_2 \cdot x_4$	$a_1 \cdot x_4$	$a_0 \cdot x_4$				
P_9	P_8	P_7	P_6	P_5	P_4	P_3	P_2	P_1	P_0

FIGURE 6.17   The partial products generated in a 5×5 multiplication.

partial product $(a_i \cdot x_j)$, one bit of the previously accumulated partial product, and a carry-in bit. A block diagram of a 5×5 array multiplier for unsigned numbers is depicted in Figure 6.18. In the first four rows there is no horizontal carry propagation. In other words, a carry-save type addition is performed in these rows, and the accumulated partial product consists of intermediate sum and carry bits. Only in the last row is a horizontal carry propagation allowed. The last row of cells in this figure is a ripple-carry adder that can be replaced by a fast two-operand adder (e.g., carry-look-ahead adder) if a shorter overall execution time is desired.

The array multiplier in Figure 6.18 has to be modified in order to allow multiplication of signed numbers in two's complement notation, since product bits like $a_4 \cdot x_0$ and $a_0 \cdot x_4$ have a negative weight and should be subtracted rather than added. One way to handle the eight negatively weighted partial product bits properly, in a 5×5 bit multiplication, is depicted in Figure 6.19. Bits with negative weight are marked with a small circle instead of an arrow. Such bits have to be subtracted instead of being added. The cells with three positive inputs are ordinary FAs and are marked by I in the figure. The cells with a single negative input and two positive inputs are marked by II. The sum of the three inputs of a type II cell can vary from -1 to 2. This requires the diagonal output c to have a weight of $+2$, and the vertical output s to have a weight of -1. The arithmetic operation of a type II cell is described by the equation

$$x + y - z = 2c - s. \qquad (6.3)$$

The values of the s and c outputs are given by

$$s = (x + y - z) \bmod 2 \quad \text{and} \quad c = \frac{(x + y - z) + s}{2}. \qquad (6.4)$$

Cells with two negative inputs and one positive input are marked by II'. The sum of their inputs can vary from -2 to 1. Hence, their c output should have a weight of -2 and their s output should have a weight of $+1$. Finally, a cell

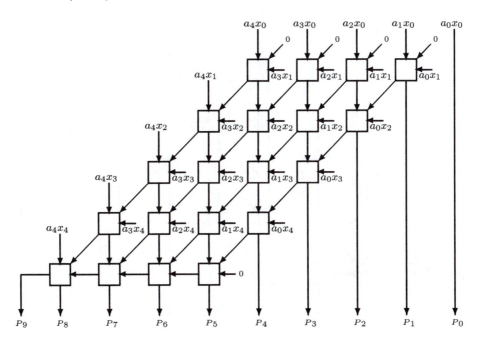

FIGURE 6.18   An array multiplier for unsigned numbers.

with all its inputs negative is marked by I' (see Figure 6.23) and has negatively weighted c and s outputs. This cell counts the number of (-1)'s at its inputs and represents this number through the c and s outputs. Logically, its operation is the same as that of type I cell and, therefore, their gate implementations are identical. This explains the reason for marking them I and I'. Similarly, the gate implementations of type II and type II' cells are identical.

Another approach to the design of an array multiplier for two's complement operands is to employ Booth's algorithm. A multiplier based on this algorithm consists of n rows of basic cells, where n is the number of multiplier bits. Each row is capable of either adding or subtracting a properly aligned multiplicand to the previously accumulated partial product. The cells in row i perform an add, subtract or transfer-only operation, depending on the value of x_i and the appropriate reference bit. Such a multiplier is shown in Figure 6.20 for four bit operands. The basic cell in this multiplier is a controlled add/subtract/shift (CASS) circuit, depicted in Figure 6.20(a) [12]. The H and D signals are control signals indicating the type of operation to be performed by the corresponding row of CASS cells. If H is 0, no arithmetic operation is done, and therefore the new partial product bit, denoted by P_{out}, is equal to the previous one, P_{in}.

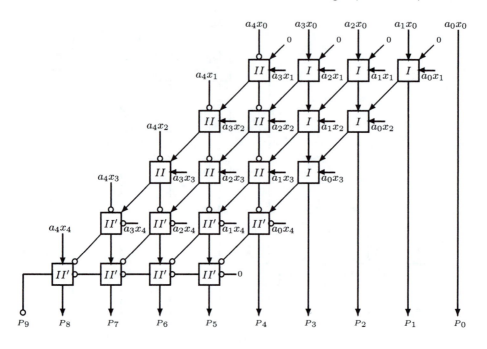

FIGURE 6.19   An array multiplier for two's complement numbers.

If $H = 1$, an arithmetic operation is performed, generating a new P_{out}. The type of arithmetic operation is indicated by the D signal. If $D = 0$ then the multiplicand bit, denoted by a, is added to P_{in} with c_{in} as an incoming carry bit from the adjacent cell to the right. The cell then generates P_{out} and c_{out} as the outgoing carry to the next cell on the left. If $D = 1$ then the multiplicand bit, a, is subtracted from P_{in}, with c_{in} as an incoming borrow and c_{out} as the outgoing borrow. Thus, the logic equations for P_{out} and c_{out} are [12]

$$P_{out} = P_{in} \oplus (a \cdot H) \oplus (c_{in} \cdot H),$$

$$c_{out} = (P_{in} \oplus D) \cdot (a + c_{in}) + a \cdot c_{in}.$$

An alternate approach to the design of a CASS cell is as a combination of a multiplexer (selecting among 0, $+a$ and $-a$) and an FA. The control signals H and D for row i are generated by a *CTRL* circuit (shown in Figure 6.20(b)) based on the multiplier bit x_i and the reference bit x_{i-1}, following the rules of Booth's algorithm from Table 6.1. The first row corresponds to the most significant bit of the multiplier. Hence, the resulting partial product needs to be shifted to the left before we add to it (or subtract from it) the next multiple of the multiplicand. To achieve this, a new cell with input $P_{in} = 0$ is added (at the right end) to the

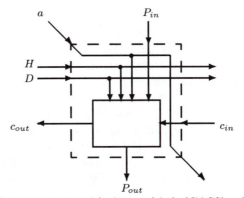

(a) Controlled add/subtract/shift (CASS) cell.

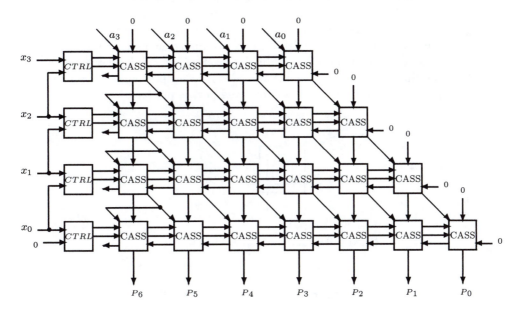

(b)

FIGURE 6.20   A Booth's algorithm array multiplier.

second row, and to each row afterward. Since the number of bits in the partial product increases by one in each row, we need to expand the multiplicand before adding it to (or subtracting it from) the partial product. This is accomplished by replicating the sign bit of the multiplicand as shown in Figure 6.20(b).

Note that we cannot take advantage of strings of 0's or 1's in this implementation, since we cannot eliminate or skip rows. Thus, the only advantage in this implementation is the ability to multiply negative numbers in two's complement with no need for any correction step. Also note that the operation in row i need not be delayed until all the upper $(i-1)$ rows have completed their operation. Thus, the least significant bit of the product, P_0, will be generated after one CASS cell delay (in addition to the delay of a CTRL circuit), P_1 will be generated after two CASS cell delays, and the most significant bit, P_{2n-2}, will be generated after $(2n-1)$ CASS cell delays.

In a similar way we can implement higher-radix multiplication schemes, which require less rows in the array by employing, for example, the radix-4 algorithms shown in Tables 6.3 and 6.5 or similar radix-8 algorithms. These, too, can handle negative multipliers in two's complement representation. The building block of such multipliers is a multiplexer-adder circuit that selects the correct multiple of the multiplicand A and adds it to the previously accumulated partial product to produce a new accumulated partial product.

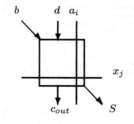

Latched full adder with an AND gate.

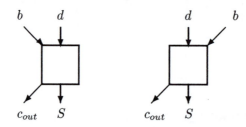

Latched half adders.

FIGURE 6.21   The basic units of the pipelined array multiplier (16).

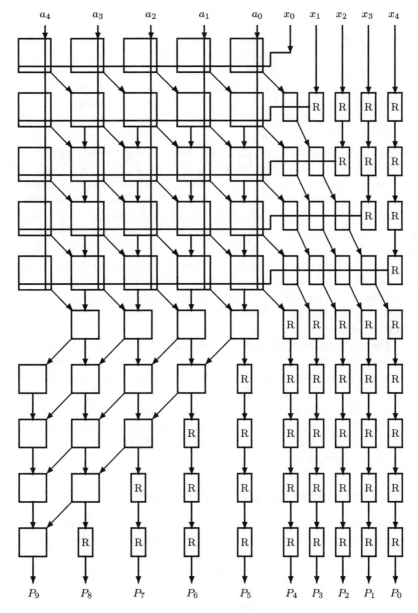

FIGURE 6.22   A pipelined 5 × 5 array multiplier for unsigned numbers.

An important characteristic of array multipliers is that they allow a pipelining mode of operation, where the execution of separate multiplications overlaps. If this mode of operation is desired, the long delay associated with the carry-propagating addition performed in the last row of the array (e.g., see Figure 6.18) should be minimized, since it determines the throughput of the pipeline. This can be achieved by replacing the CPA with several additional rows that, like the first rows in the array, allow a carry propagation of only one position between any two consecutive rows. Five such rows are needed in the 5×5 array multiplier for unsigned numbers in Figure 6.18, with 4, 4, 3, 2, and 1 cells, respectively. These rows are shown in a pipelined version of the 5×5 array multiplier, depicted in Figure 6.22. The basic cells employed in this multiplier are shown in Figure 6.21 [16]. The FA in Figure 6.21 includes an AND gate that generates the product bit $a_i x_j$. This product bit is added to the incoming bits b and d to produce the output bits S and c_{out}. The modified FA also propagates the bits a_i and x_j to neighboring cells. The two versions of the HA in the same figure are used in the bottom five rows where the cells add only two input bits each.

In order to support pipelining, all cells in the array must include latches, so that each row can handle a separate multiplier-multiplicand pair. Also, registers are needed to propagate the multiplier bits to their destination, and to propagate the product bits that have been completed, which is done in parallel with the generation of new product bits.

Up to 10 consecutive multiply operations can be executed simultaneously in the multiplier depicted in Figure 6.22. The maximum rate at which multiply operations can be completed is determined by the delay associated with the modified FA, including the latches. This rate might be, in practice, too high to be used as the clock rate of the circuit. However, other implementations of the 5×5 pipelined multiplier with a lower rate are possible. For example, two rows can be combined to form a single pipeline stage with a lower rate but with fewer latches (less circuitry overall).

6.7 OPTIMALITY OF MULTIPLIER IMPLEMENTATIONS

Bounds on the performance of algorithms for multiplication have been derived in a way similar to that of the bounds for addition that were described in Section 5.4. It is interesting to note that the theoretical bounds for multiplication are similar to those for addition, although, in practice, multiplication is more time-consuming than addition. Thus, if we adopt the idealized model, which assumes that all circuits are implemented using (f, r) gates, the execution time of a multiply circuit for two operands with n bits each must satisfy

$$T_{mult} \geq \lceil \log_f 2n \rceil. \qquad (6.5)$$

Also, if the residue number system is employed, smaller circuits with fewer inputs are required and, consequently, the lower bound is

$$T_{mult} \geq \lceil \log_f 2m \rceil, \tag{6.6}$$

where m is the number of digits that are needed to represent the largest modulus in the residue number system, as explained in Chapter 11, and usually $m \ll n$.

When searching for an optimal implementation of a multiplier in the conventional binary number system, we need to compare the performance (execution time) and implementation costs (e.g., regularity of the design, total area, etc.) of the previously described algorithms for multiplication. When both execution time and implementation cost, say, area, need to be taken into account, an objective function like $A \cdot T$ can be used, where A denotes the area and T denotes the execution time. A more general form of an objective function is $A \cdot T^\alpha$, where α can be either smaller or larger than 1.

In what follows we compare several multipliers, some of which were presented in previous sections. The simple array multiplier depicted in Figure 6.18 has a very regular structure. It can be implemented easily as a rectangular-shaped array, with no waste of chip area. The n least significant bits of the final product are then produced on the right side of the rectangle, while the n most significant product bits are the outputs of the bottom row of the rectangle, which constitutes a CPA. Although this implementation is highly regular and its design and layout are very simple, it possesses two major drawbacks. First, it requires a very large area, proportional to n^2, since it contains about n^2 FAs and AND gates. Second, it has a long execution time T of about $2n \cdot \Delta_{FA}$ (Δ_{FA} is the delay of an FA). More precisely, T consists of $(n-1)\Delta_{FA}$ for the first $(n-1)$ rows and an additional $(n-1)\Delta_{FA}$ for the CPA if implemented as a ripple-carry adder, as shown in Figure 6.18. Thus, an objective function of the form $A \cdot T$ is directly proportional to n^3.

If a highly pipelined version of this array multiplier is desired, the required area increases even further (since the CPA must be replaced) as does the latency of a single multiply operation. However, the resulting pipeline period, which determines the pipelining rate, is shorter.

An implementation of the Booth's algorithm array multiplier, depicted in Figure 6.20, offers no advantage over the previous multiplier when performance and area are considered, since the area A is of the order of n^2 and T is linear in n. The modified radix-4 Booth algorithm (see Table 6.3) can potentially result in a better implementation, since it requires only $n/2$ rows of cells. This reduction in the number of rows could, in principle, reduce the delay (T) and the implementation cost (A) by a factor of two, decreasing the objective function $A \cdot T$ to a fourth of its previous value. However, a more detailed examination of the design reveals that the actual delay and area gains are less than expected. The recoding logic and, more importantly, the partial product selectors, add complexity to the circuit and result in a larger number of interconnections and a longer delay per row. Also, since the relative shift between any two adjacent rows is two bit positions, we must allow the carry to propagate horizontally in these bit positions. This can be achieved either locally or at the last row of the array multiplier. After that, a carry propagation through $(2n - 1)$ bits (instead of $n - 1$) is required [18]. The exact overall reduction in the objective function depends on the details of the design and the technology used.

Similar problems are encountered when implementing the radix-8 modified Booth's algorithm in the form of an array multiplier. In addition, the partial product $3A$ should be precalculated. Consequently, the reduction in delay and area may be far less than the expected factor of $1/3$. Still, the implementation of the radix-8 algorithm might be cost-effective in certain technologies and design styles.

Irrespective of the way partial products are generated, they can be accumulated either through a cascade interconnection (as in Figures 6.18 and 6.20) or through a tree structure (e.g., a CSA tree, as in Section 5.11, or some variation of it, as in Section 6.4). The number of levels in a CSA tree for k partial products is of the order of $\log k$ rather than being linear in k (as in a cascade interconnection), resulting in a much shorter execution time (the number of partial products, k, can be n, $n/2$, $n/3$, etc; where n is the number of bits). However, CSA tree structures have irregular interconnects, making it difficult to find an area-efficient layout with a rectangular shape. Moreover, an overall width of $2n$ is required in most cases. This may result in a multiplier area of the order of $2n \log k$. The objective function $A \cdot T$ may, consequently, increase as $2n \log^2 k$.

The balanced delay tree in Figure 6.13 has a more regular structure. The increments in the number of operands in the balanced delay tree are $3, 3, 5, 7, 9 \cdots$. The sum of the elements in this series is of the order of j^2, where j is the number of elements in the series. The number of levels, which determines the overall delay, increases linearly with j. As a result, the overall delay of a balanced delay tree is of the order of $\sqrt{k}$, where $k = j^2$ is the number of operands. This needs to be compared to $\log k$, which is the number of levels in the complete binary tree. The detailed proof is left to the reader as an exercise. One should be aware that general expressions for the complexity of either the execution time or the area, like the ones above, have theoretical importance, but only limited practical significance. For any given technology, a more detailed examination of the alternative designs is necessary before final conclusions can be drawn.

6.8 EXERCISES

6.1. Show that Booth's algorithm can be used to convert a number in two's complement representation to its SD representation.

6.2. Prove that no correction step is needed when using the multiplication algorithm in Table 6.3 with a negative multiplier represented in two's complement. Repeat this for the algorithm in Table 6.5 with a sign bit extension.

6.3. Verify that the new partial product in the radix-4 modified Booth algorithm is $(x_{i-1} + x_{i-2} - 2x_i) \cdot A$ for odd values of i. Use this expression to formally prove the correctness of the algorithm.

6.4. Write down the rules for a radix-8 modified Booth's algorithm or, in other words, a 3-bit version of the algorithm in Table 6.3.

6.5. (a) Verify that the 74261 chip, which is called a 2-bit by 4-bit parallel multiplexer, implements the algorithm in Table 6.3.
(b) How many such chips are needed to construct a 12×12 bit two's complement multiplier? Show how these chips should be interconnected.
(c) Explain how the $\overline{Q_4}$ output signals of the 74261 chip are used to generate the sign bit of the partial product.
(d) What type of carry-save adder is needed?

6.6. In case 101 of Table 6.6 we need two forced carries into the adder. To avoid this, we may force x_0 to 0 if it equals 1 and set the initial partial product to be $+A$ instead of $-A$. Show that the correct partial product is always obtained.

6.7. Design a $3n \times 3n$ bit multiplier out of $n \times n$ bit multipliers. Find the number of $n \times n$ bit multipliers that are needed and show how the partial products should be aligned. What type of counters are needed to add the partial products? Can (5,5,4) counters be useful?

6.8. Write the truth table of a type II cell used in two's complement array multipliers and obtain the Boolean equations for the c and s outputs. Repeat this for type II' cells.

6.9. Can the four cells in the last row in Figure 6.19 be made into type II by defining the rightmost zero carry as having a positive weight?

6.10. The idea behind the array multiplier in Figure 6.19 was first proposed by Pezaris [19], who has shown a slightly different organization of the multiplier, as depicted in Figure 6.23. Explain why the P_4 output in Figure 6.23 is connected to the cell on its left side.

6.11. Design an array multiplier for two 5-bit nega-binary operands; for example, $X = \sum_{i=0}^{4} x_i(-2)^i$. What is the range of each operand and of the product? Draw the multiplier, indicate how many different types of 1-bit cells are needed, and give the truth table for each type.

6.12. In this question you are asked to estimate the execution time of an array multiplier like the one shown in Figure 6.18. Denote by Δ_s and Δ_c the delays associated with the sum and carry outputs of the basic cells, respectively, and assume that they satisfy $\Delta_s > \Delta_c$. Find the critical path in the array multiplier assuming all product bits $a_i x_j$ are available simultaneously. Estimate the execution time of a $n \times n$ bits multiplication. Can you suggest ways to speed up the operation?

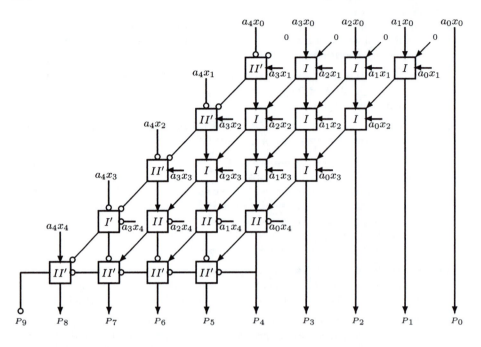

FIGURE 6.23   The array multiplier for two's complement numbers suggested in (19).

6.13. Prove that the arrangement of partial product bits shown in Figure 6.24 produces the correct product of two 5-bit two's complement operands, where $\overline{x}_i = 1 - x_i$ and similarly $\overline{a}_i = 1 - a_i$. This arrangement was suggested by Baugh and Wooley [1]. Compare this multiplier to the two's complement array multipliers shown in Figures 6.19 and 6.20, considering the amount of hardware and execution time.

6.14. The Booth's algorithm multiplier in Figure 6.20 starts with the most significant bit of the multiplier. Redesign the multiplier starting with the least significant bit of X. Compare the execution time and the required hardware of the two alternatives.

6.15. Show a block diagram of a 6×6 bit two's complement multiplier constructed out of multiplexer-adder circuits based on the radix-4 modified Booth's algorithm in Table 6.3.

6.16. Prove that the delay of the balanced delay tree shown in Figure 6.13 is proportional to $\sqrt{k}$, where k is the number of operands.

6.17. Explain why the HAs in the leftmost column of the array multiplier in Figure 6.22 have no carry output.

6.18. Verify that the implementation in Figure 6.12 corresponds to the setting $a = b = c = 1$ and $d = e = f = 0$ of the variables in Table 6.7.

P_9	P_8	P_7	P_6	P_5	P_4	P_3	P_2	P_1	P_0
					a_4	a_3	a_2	a_1	a_0
				$\times$	x_4	x_3	x_2	x_1	x_0
					$a_4\cdot\overline{x}_0$	$a_3\cdot x_0$	$a_2\cdot x_0$	$a_1\cdot x_0$	$a_0\cdot x_0$
				$a_4\cdot\overline{x}_1$	$a_3\cdot x_1$	$a_2\cdot x_1$	$a_1\cdot x_1$	$a_0\cdot x_1$	
			$a_4\cdot\overline{x}_2$	$a_3\cdot x_2$	$a_2\cdot x_2$	$a_1\cdot x_2$	$a_0\cdot x_2$		
		$a_4\cdot\overline{x}_3$	$a_3\cdot x_3$	$a_2\cdot x_3$	$a_1\cdot x_3$	$a_0\cdot x_3$			
	$a_4\cdot x_4$	$\overline{a}_3\cdot x_4$	$\overline{a}_2\cdot x_4$	$\overline{a}_1\cdot x_4$	$\overline{a}_0\cdot x_4$				
	$\overline{a}_4$	0	0	0	a_4				
1	$\overline{x}_4$	0	0	0	x_4				
P_9	P_8	P_7	P_6	P_5	P_4	P_3	P_2	P_1	P_0

FIGURE 6.24   Partial products for a 5×5 two's complement multiplication (1).

6.19. Find the values of a, b, c, d, e and f in Table 6.7 which will yield an expression with the smallest number of literals (a literal is any appearance of either x_i or $\overline{x_i}$) for c_{out}.

6.20. Prove that the following modification (see Figure 6.25) of the arrangement of partial products (for two's complement operands) suggested in [1], produces the correct final product. Compare this arrangement to the original one shown in Figure 6.24.

P_9	P_8	P_7	P_6	P_5	P_4	P_3	P_2	P_1	P_0
				1	$\overline{a_4\cdot x_0}$	$a_3\cdot x_0$	$a_2\cdot x_0$	$a_1\cdot x_0$	$a_0\cdot x_0$
				$\overline{a_4\cdot x_1}$	$a_3\cdot x_1$	$a_2\cdot x_1$	$a_1\cdot x_1$	$a_0\cdot x_1$	
			$\overline{a_4\cdot x_2}$	$a_3\cdot x_2$	$a_2\cdot x_2$	$a_1\cdot x_2$	$a_0\cdot x_2$		
		$\overline{a_4\cdot x_3}$	$a_3\cdot x_3$	$a_2\cdot x_3$	$a_1\cdot x_3$	$a_0\cdot x_3$			
1	$a_4\cdot x_4$	$\overline{a_3\cdot x_4}$	$\overline{a_2\cdot x_4}$	$\overline{a_1\cdot x_4}$	$\overline{a_0\cdot x_4}$				
P_9	P_8	P_7	P_6	P_5	P_4	P_3	P_2	P_1	P_0

FIGURE 6.25   Modified array of partial products for a 5×5 two's complement multiplication.

6.9 REFERENCES

[1] C.R. BAUGH and B.A. WOOLEY, "A two's complement parallel array multiplication algorithm," *IEEE Trans. on Computers, C-22* (Dec. 1973), 1045-1047.

[2] K. C. BICKERSTAFF, M. J. SCHULTE and E. E. SWARTZLANDER, "Parallel reduced area multipliers," *Journal of VLSI Signal Processing, 9,* (1995), 181-191.

[3] A.D. BOOTH, "A signed binary multiplication technique," *Quart. J. Mech. Appl. Math., 4,* Part 2, 1951, 236-240.

[4] L. DADDA, "Some schemes for parallel multipliers," *Alta Frequenza, 34* (March 1965), 346-356.

[5] L. DADDA, "On parallel digital multipliers," *Alta Frequenza, 45* (1976), 574-580.

[6] J. DEVERELL, "Pipeline iterative arithmetic arrays," *IEEE Trans. on Computers, C-24* (March 1975), 317-322.

[7] M. J. FLYNN AND S. F. OBERMAN, *Advanced computer arithmetic design,* Wiley, New York, 2001.

[8] J.A. GIBSON and R.W. GIBBARD, "Synthesis and comparison of two's complement parallel multipliers," *IEEE Trans. on Computers, C-24* (Oct. 1975), 1020-1027.

[9] A. HABIBI and P.A. WINTZ, "Fast multipliers," *IEEE Trans. on Computers, C-19* (Feb. 1970), 153-157.

[10] Y. HARATA *et al.,* "A high speed multiplier using a redundant binary adder tree," *IEEE J. of Solid-State Circuits, SC-22* (Feb. 1987), 28-33.

[11] O. L. MACSORLEY, "High-speed arithmetic in binary computers," *Proc. of IRE, 49* (Jan. 1961), 67-91.

[12] J.C. MAJITHIA and R. KITA, "An iterative array for multiplication of signed binary numbers," *IEEE Trans. on Computers, C-20* (Feb. 1971), 214-216.

[13] M. MEHTA, V. PARMAR and E. SWARTZLANDER, "High-speed multiplier design using multi-input counter and compressor circuits," *Proc. 10th Symp. on Computer Arithmetic* (1991), 43-50.

[14] R. MONTOYE, E. HOKENEK and S.L. RUNYON, "Design of the IBM RISC system/600 floating-point unit," *IBM Journal of Research and Development, 34* (January 1990), 59-67.

[15] Z.J. MOU and F. JUTAND, "Overturned-stairs adder trees and multiplier design," *IEEE Trans. on Computers, 41* (August 1992), 940-948.

[16] T.G. NOLL *et al.,* "A pipelined 330-MHz multiplier," *IEEE Journal of Solid-State Circuits, SC-21* (June 1986), 411-416.

[17] V. G. OKLOBDZIJA and D. WILLEGER, "Improving multiplier design by using improved column compression tree and optimized final adder in CMOS technology," *IEEE Trans. on VLSI systems, 3* (June 1995), 292-301.

[18] V. PENG, S. SAMUDRALA and M. GAVRIELOV, "On the implementation of shifters, multipliers and dividers in VLSI floating-point units," *Proc. of 8th Symp. on Computer Arithmetic* (May 1987), 95-102.

[19] S.D. PEZARIS, "A 40ns 17-bit by 17-bit array multiplier," *IEEE Trans. on Computers, C-20* (April 1971), 442-447.

[20] G. W. REITWIESNER, "Binary arithmetic," in *Advances in computers,* vol. 1, F. L. Alt, (Editor), Academic, New York, 1960, pp. 231-308.

[21] L.P. RUBINFIELD, "A proof of the modified Booth's algorithm for multiplication," *IEEE Trans. on Computers, C-24* (Oct. 1975), 1014-1015.

[22] M.R. SANTORO and M.A. HOROWITZ, "SPIM: A pipelined 64×64 iterative multiplier," *IEEE Journal of Solid-State Circuits, 24* (April 1989), 487-493.

[23] P.F. STELLING, C.U. MARTEL, V.G. OKLOBDZIJA and R. RAVI, "Optimal circuits for parallel multipliers," *IEEE Trans. on Computers, 47* (March 1998) 273-285.

[24] D. ZURAS and W.H. MCALLISTER, "Balanced delay trees and combinatorial division in VLSI," *IEEE Journal of Solid-State Circuits, SC-21* (Oct. 1986), 814-819.

[25] R.K. YU and G.B. ZYNER, "167 MHz radix-4 floating point multiplier," *Proc. of the 12th Symp. on Computer Arithmetic* (July 1995), 149-154.

7

FAST DIVISION

There are two different approaches to the development of algorithms for high-speed division. The more conventional approach uses add/subtract and shift operations, while the second relies on multiplication. The operation count in the first approach is linearly proportional to the word size, n. The number of steps in the second approach is logarithmically proportional to n, but each individual step is more complex. The first approach is discussed in this chapter while the second is presented in Chapter 8.

7.1 SRT DIVISION

The most well known division algorithm of the first type is the SRT division, named after Sweeney, Robertson, and Tocher ([11], [15], [19]), each of whom developed it independently at around the same time. The motivation behind the SRT algorithm was an attempt to speed up the nonrestoring division (which consists of n add/subtract operations and is presented in Chapter 3) by allowing 0 to be a quotient digit for which no add/subtract operation is needed. In principle, we can change the rule for selecting the quotient digit in the nonrestoring division to

$$q_i = \begin{cases} 1 & \text{if } 2r_{i-1} \geq D \\ 0 & \text{if } -D \leq 2r_{i-1} < D \\ \bar{1} & \text{if } 2r_{i-1} < -D \end{cases} \qquad (7.1)$$

and the corresponding new remainder is

$$r_i = 2r_{i-1} - q_i \cdot D. \qquad (7.2)$$

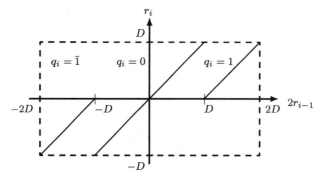

FIGURE 7.1   Nonrestoring division with $q_i = 0$.

This modified nonrestoring division is diagrammed in Figure 7.1. The difficulty with this new selection rule is that a full comparison of $2r_{i-1}$ with either D or $-D$ is required. If we restrict D to be a normalized fraction satisfying $\frac{1}{2} \leq |D| < 1$, we may reduce the region of $2r_{i-1}$, for which $q_i = 0$, as follows:

$$-D \leq -\frac{1}{2} \leq 2r_{i-1} < \frac{1}{2} \leq D \tag{7.3}$$

The advantage of this is that now we can compare the partial remainder $2r_{i-1}$ to either $1/2$ or $-1/2$, instead of D or $-D$. A binary fraction is larger than or equal to $1/2$ if, and only if, it starts with 0.1. Similarly, a binary fraction is smaller than $-1/2$ if and only if it starts with 1.0 (in two's complement representation). Consequently, only two bits of $2r_{i-1}$ have to be examined, instead of a full-length comparison between $2r_{i-1}$ and D. In some cases (e.g., when the dividend X is larger than $1/2$) the shifted partial remainder needs an integer bit in addition to the sign bit, and thus, three bits of $2r_{i-1}$ must be examined. The rule for selecting the quotient digit becomes

$$q_i = \begin{cases} 1 & \text{if } 2r_{i-1} \geq 1/2 \\ 0 & \text{if } -1/2 \leq 2r_{i-1} < 1/2 \\ \bar{1} & \text{if } 2r_{i-1} < -1/2. \end{cases} \tag{7.4}$$

The resulting algorithm is called SRT division, and it can be diagrammed as shown in Figure 7.2. This diagram shows the quotient digits that must be selected in order to satisfy the condition $|r_i| \leq |D|$, guaranteeing the convergence of the division procedure with a final remainder smaller than $|D|$.

The SRT division process starts off with a normalized divisor and has the effect of normalizing the partial remainder by shifting over leading 0's if it is positive, and leading 1's if it is negative. For example, if $2r_{i-1} = 0.001xxxx$ (where x is any binary digit), then $2r_{i-1} < 1/2$ and we set q_i to 0, obtaining

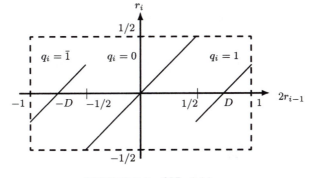

FIGURE 7.2   SRT division.

$2r_i = 0.01xxxx$ and so on. Similarly, if $2r_{i-1} = 1.110xxxx$ then $2r_{i-1} > -1/2$ and we set $q_i = 0$ obtaining $2r_i = 1.10xxxx$ and so on. We say, therefore, that SRT division is nonrestoring division with a normalized divisor and remainder.

SRT division, as nonrestoring division, can be extended to include negative divisors in two's complement. The selection rule for q_i then becomes

$$q_i = \begin{cases} 0 & \text{if} \quad |2r_{i-1}| < 1/2 \\ 1 & \text{if} \quad |2r_{i-1}| \geq 1/2 \ \& \ r_{i-1} \text{ and } D \text{ have the same sign} \\ \bar{1} & \text{if} \quad |2r_{i-1}| \geq 1/2 \ \& \ r_{i-1} \text{ and } D \text{ have opposite signs.} \end{cases}$$

$$(7.5)$$

Example 7.1

Let the dividend X be equal to $(0.0101)_2 = 5/16$ and the divisor D be $(0.1100)_2 = 3/4$. Applying the SRT algorithm yields

$r_0 = X$		0	.0	1	0	1		
$2r_0$		0	.1	0	1	0	$\geq 1/2$ set $q_1 = 1$	
Add $-D$	+	1	.0	1	0	0		
r_1		1	.1	1	1	0		
$2r_1 = r_2$		1	.1	1	0	0	$\geq -1/2$ set $q_2 = 0$	
$2r_2 = r_3$		1	.1	0	0	0	$\geq -1/2$ set $q_3 = 0$	
$2r_3$		1	.0	0	0	0	$< -1/2$ set $q_4 = \bar{1}$	
Add D	+	0	.1	1	0	0		
r_4		1	.1	1	0	0	negative remainder & positive X	
Add D	+	0	.1	1	0	0	correction	
r_4		0	.1	0	0	0	corrected final remainder	

The quotient generated before the correction is $Q = 0.100\bar{1}$. This is a minimal representation of $Q = 0.0111$ in SD form. In other words, a minimal number of add/subtract operations is performed. After correction, Q becomes $0.0111 - ulp = 0.0110_2 = 3/8$, and the final remainder is $1/2 \cdot 2^{-4} = 1/32$.   □

Example 7.2

Let $X = (0.00111111)_2 = 63/256$ and $D = (0.1001)_2 = 9/16$.

$r_0 = X$		0	.0	0	1	1	1	1	1	1
$2r_0$		0	.0	1	1	1	1	1	1	0	$< 1/2$ set $q_1 = 0$
$2r_1$		0	.1	1	1	1	1	1	0	0	$\geq 1/2$ set $q_2 = 1$
Add $-D$	+	1	.0	1	1	1				
r_2		0	.0	1	1	0	1	1	0	0
$2r_2$		0	.1	1	0	1	1	0	0	0	$\geq 1/2$ set $q_3 = 1$
Add $-D$	+	1	.0	1	1	1				
r_3		0	.0	1	0	0	1	0	0	0
$2r_3$		0	.1	0	0	1	0	0	0	0	$\geq 1/2$ set $q_4 = 1$
Add $-D$	+	1	.0	1	1	1				
r_4		0	.0	0	0	0	0	0	0	0	zero final remainder

$Q = 0.0111_2 = 7/16$. This is not a minimal representation of the quotient in SD form.   □

Based on the last example, we may conclude that it is possible to further reduce the number of add/subtract operations. Simulations and statistical analysis studying the efficiency of the SRT method have been performed [9], and the conclusions were:

1. The average "shift" in the SRT method is 2.67, meaning that for a dividend of length n we need, on the average, $n/2.67$ operations. For example, for $n = 24$, on the average, $24/2.67 = 8.9 \approx 9$ operations are required.

2. The actual number of operations needed depends upon the divisor D. The smallest number is achieved when $17/28 \leq D \leq 3/4$ (or, approximately when $3/5 \leq D \leq 3/4$), with an average shift of 3.

Hence, in order to reduce the number of add/subtract operations, we should modify the SRT method when the divisor happens to be out of the optimal range ($3/5 \leq D \leq 3/4$). Two ways of achieving this are described below:

1. Examine the possibility of using a multiple of D like $2D$ if D is too small, or $D/2$ if D is too large, in some of the steps during the division. Notice

that subtracting $2D$ $(D/2)$ instead of D is equivalent to performing the subtraction one position earlier (later).

2. Change the comparison constant $K = 1/2$ if D is outside the optimal range. Such a change is allowed because the ratio D/K is what really matters, since we compare the partial remainder to K, not D.

The idea behind scheme (1) is that whenever D is small we may end up generating a sequence of 1's in the quotient one bit at a time, requiring a subtract operation per each bit, as in the last example. In such cases, subtracting $2D$ instead of D (which is equivalent to subtracting D in the previous step) might generate a negative partial remainder, allowing us to generate a sequence of 0's as quotient bits while normalizing the partial remainder.

Example 7.3

Repeating the previous example we obtain

$r_0 = X$		0	.0	0	1	1	1	1	1	1	
$2r_0$		0	.0	1	1	1	1	1	1	0	$< 1/2$ set $q_1 = 0$
$2r_1$	0	0	.1	1	1	1	1	1	0	0	subtract $2D$
Add $-2D$ +	1	0	.1	1	1						instead of D
r_2	1	1	.1	1	0	1	1	1	0	0	set $q_1 = 1$ and $q_2 = 0$
$2r_2$		1	.1	0	1	1	1	0	0	0	set $q_3 = 0$
$2r_3$		1	.0	1	1	1	0	0	0	0	$\leq -1/2$ set $q_4 = \bar{1}$
Add D +		0	.1	0	0	1					
r_4		0	.0	0	0	0	0	0	0	0	zero final remainder

$Q = 0.100\bar{1}_2 = 7/16$ and this is a minimal representation of the quotient in SD form. □

If D is large, a single 0 within a sequence of 1's in the quotient may result in two consecutive add/subtract operations, instead of one. Performing an addition of $D/2$ instead of D for the last 1 before the single 0 (which is equivalent to performing the addition one position later) may generate a negative partial remainder that will allow us to properly handle the single 0, and then continue normalizing the partial remainder until the end of the sequence of 1's is reached.

Example 7.4

Let $X = (0.01100)_2 = 3/8$ and $D = (0.11101)_2 = 29/32$. The correct 5-bit quotient is $Q = 0.01101_2 = 13/32$. Applying the basic SRT algorithm results in $Q = 0.10\bar{1}1\bar{1}$, and the single 0 within the group of 1's in Q is not handled in the most efficient way. If we use the multiple $D/2$ we obtain

$r_0 = X$		0	.0	1	1	0	0	
$2r_0$		0	.1	1	0	0	0	$\geq 1/2$ set $q_1 = 1$
Add $-D$	+	1	.0	0	0	1	1	
r_1		1	.1	1	0	1	1	
$2r_1$		1	.1	0	1	1	0	set $q_2 = 0$
$2r_2$		1	.0	1	1	0	0	0
Add $D/2$	+	0	.0	1	1	1	0	1
r_3		1	.1	1	0	1	0	1
$2r_3$		1	.1	0	1	0	1	
$2r_4$		1	.0	1	0	1	0	
Add D	+	0	.1	1	1	0	1	
r_5		0	.0	0	1	1	1	final remainder = $7/32 \cdot 2^{-5}$

Notes column (right of table):
- $2r_2$: add $D/2$ $q_3 = \bar{1}$)
- Add $D/2$: instead of D
- r_3: set $q_3 = 0$ and
- $2r_3$: $q_4 = \bar{1}$
- $2r_4$: $\leq -1/2$ set $q_5 = \bar{1}$

$Q = 0.100\bar{1}\bar{1}_2 = 13/32$; i.e., the single 0 is handled properly.   □

To implement this scheme, two adders are needed. One adder will always add or subtract D, while the other will add/subtract $2D$ if D is too small (i.e., D starts with 0.10 in its true form) or add/subtract $D/2$ if D is too large (i.e., D starts with 0.11 in its true form) [11]. The output of the primary adder is normally used, unless the output of the alternate adder results in a larger normalizing shift.

The idea of using multiples of D can be extended to the use of $3D/2$ and $3D/4$ in addition to D itself. These provide an even higher overall average shift (of about 3.7), but require a more complex implementation [11].

Scheme (2) is based on the fact that for $K = 1/2$, the ratio D/K in the optimal range $3/5 \leq D \leq 3/4$ is

$$6/5 \leq \frac{D}{K} = \frac{D}{1/2} \leq 3/2 \qquad \text{or} \qquad \frac{6}{5} K \leq D \leq \frac{3}{2} K. \qquad (7.6)$$

If the given D is not in the optimal range for $K = 1/2$, we can choose a different comparison constant K. Consequently, for different ranges of D there are different values of K.

A numerical search for K [13] has shown that satisfactory results can be obtained if the region $1/2 \leq |D| < 1$ is divided into five (not equally sized) subregions, each having a different comparison constant K_i, as depicted in Figure 7.3. Note that four bits of the divisor have to be examined in order to select the comparison constant, which in turn has only four bits to be compared to the four most significant bits of the remainder. The determination of the subregions for the divisor and the corresponding comparison constants has to be done through a numerical search. This is because both should be binary fractions, with a small number of bits in order to simplify the resulting division algorithm.

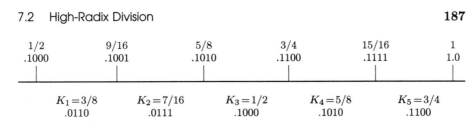

FIGURE 7.3 The values of the comparison constant for the five divisor subregions.

Example 7.5

We repeat the division in Example 7.2 with $X = (0.00111111)_2 = 63/256$ and $D = (0.1001)_2 = 9/16$. The appropriate comparison constant for this divisor is $K_2 = 7/16 = 0.0111_2$ (see Figure 7.3). If the remainder is negative, it should be compared to the two's complement of K_2, which is 1.1001_2.

$r_0 = X$		0	.0	0	1	1	1	1	1	1
$2r_0$		0	.0	1	1	1	1	1	1	0	≥ 0.0111 set $q_1 = 1$
Add $-D$	$+$	1	.0	1	1	1				
r_1		1	.1	1	1	0	1	1	1	0
$2r_1 = r_2$		1	.1	1	0	1	1	1	0	0	≥ 1.1001 set $q_2 = 0$
$2r_2 = r_3$		1	.1	0	1	1	1	0	0	0	≥ 1.1001 set $q_3 = 0$
$2r_3$		1	.0	1	1	1	0	0	0	0	< 1.1001 set $q_4 = \bar{1}$
Add D	$+$	0	.1	0	0	1				
r_4		0	.0	0	0	0	0	0	0	0	zero final remainder

The quotient $Q = 0.100\bar{1} = 0.0111_2 = 7/16$ is represented in a minimal *SD* form. □

7.2 HIGH-RADIX DIVISION

The number of add/subtract operations required by the radix-2 SRT algorithm and its variations is data-dependent. Thus, an asynchronous circuit must be designed in order to take advantage of the reduced number of nonzero bits in the quotient. Consequently, attempts to increase the number of zeros in the quotient have, in the currently available technology, very limited practical significance.

The number of add/subtract operations in the division process can be reduced and still be data-independent by increasing the radix β for the process, where selecting $\beta = 2^m$ allows the generation of m quotient bits at each step. In this way, the number of steps is reduced to $\lceil n/m \rceil$. The recursive equation for the remainder is now

$$r_i = \beta \, r_{i-1} - q_i \cdot D \tag{7.7}$$

where the multiplication by $\beta = 2^m$ is achieved by shifting the remainder m bit positions to the left. The digit set for the quotient is $\{0, 1, \cdots, (\beta - 1)\}$ for the restoring division and can be as large as $\{\overline{(\beta - 1)}, \cdots, \overline{1}, 0, 1, \cdots, (\beta - 1)\}$ for the high-radix SRT division.

A radix higher than 2 can, in principle, be used for any of the previously mentioned division algorithms. For restoring division, this means that we start with the initial guess $q_i = 1$ and, if the resulting remainder $\beta r_{i-1} - D$ is positive, we increase it to $q_i = 2$ and subtract D from the temporary remainder, obtaining $\beta r_{i-1} - 2D$. The process is repeated until we reach the value $q_i = j$, for which the temporary remainder is negative. We then restore the remainder by adding D, obtaining $\beta r_{i-1} - (j-1)D$, and set $q_i = j - 1$. This sequential procedure can be very time-consuming, making its advantage over the binary algorithm questionable. It can be replaced by a parallel process if several comparison circuits comparing βr_{i-1} to multiples of the divisor, jD, are included in the division unit. The comparison circuit producing the smallest positive remainder points to the correct quotient digit. Clearly, this implementation requires a substantial hardware investment. Similar changes can be introduced into the binary nonrestoring division algorithm.

In what follows we describe the high-radix SRT algorithm. It is possible to implement a high-radix SRT division circuit that is faster than its binary version. The quotient digit q_i in such an algorithm is a signed digit in the range $\{\overline{\alpha}, \overline{\alpha - 1}, \cdots, \overline{1}, 0, 1, \cdots, \alpha\}$, where $\lceil \frac{1}{2}(\beta - 1) \rceil \leq \alpha \leq (\beta - 1)$ (see Chapter 2).

To find out the possible choices for α in the high-radix division algorithm consider the following. The quotient digit q_i is ordinarily selected so that $|r_i| < |D|$; otherwise, the next quotient digit might have to be β or larger. This guarantees the convergence of the division procedure. The above condition implies that for the maximal remainder $\beta r_{i-1} = \beta(D - ulp)$ and a positive divisor D, selecting the largest value for the quotient digit $q_i = \alpha$ should be sufficient to yield a remainder r_i in the allowable region. Therefore, the following inequality should hold:

$$r_i = \beta(D - ulp) - \alpha D \leq D - ulp \tag{7.8}$$

Dividing Equation (7.8) by D reveals that we may select for α only the maximum value $\alpha = (\beta - 1)$. It is reasonable, however, to consider division techniques for which $|r_i| \leq k|D|$, where k is a fraction, since this reduces the size of the allowable region for the partial remainder, as shown in Figure 7.4. Equation (7.8) now takes the form

$$r_i = \beta k(D - ulp) - \alpha D \leq k(D - ulp). \tag{7.9}$$

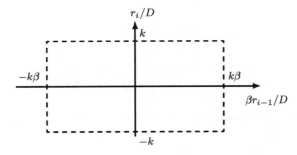

FIGURE 7.4   The allowable region for the partial remainder ($k \leq 1$).

Hence, $\alpha \geq k(\beta - 1)$, and if we want to allow the selection of any value of α in the range $\lceil \frac{1}{2}(\beta - 1) \rceil \leq \alpha \leq (\beta - 1)$, k should satisfy $1/2 \leq k \leq 1$, since $k \leq \frac{\alpha}{\beta - 1}$. The smaller the value of k, the smaller the redundancy is in the number system for the quotient.

Example 7.6

Let $\beta = 4$ and $\alpha = 2$. We set $k = \alpha/(\beta - 1) = 2/3$. For this selection, $|r_i| \leq kD = \frac{2}{3}D$, and $|\beta r_{i-1}| = |4r_{i-1}| \leq \frac{8}{3}D$, or $|\frac{r_i}{D}| \leq \frac{2}{3}$ and $|\frac{4r_{i-1}}{D}| \leq \frac{8}{3}$. The digit set for q_i is $\{\bar{2}, \bar{1}, 0, 1, 2\}$. The region in which a specific value q can be selected is given by

$$-\frac{2}{3} \leq \frac{4r_{i-1}}{D} - q \leq \frac{2}{3},$$

and consequently,

$$-\frac{2}{3} + q \leq \frac{4r_{i-1}}{D} \leq \frac{2}{3} + q.$$

For example, the value $q_i = 2$ may be selected in the region $\frac{4}{3} \leq \frac{4r_{i-1}}{D} \leq \frac{8}{3}$, since $-\frac{2}{3} \leq (\frac{4r_{i-1}}{D} - 2) \leq \frac{2}{3}$. Similarly, we may select $q_i = 1$ for $\frac{1}{3} \leq \frac{4r_{i-1}}{D} \leq \frac{5}{3}$. The different regions for selecting the quotient digits are shown in Figure 7.5. In the overlapping region, namely, $\frac{4}{3} \leq \frac{4r_{i-1}}{D} \leq \frac{5}{3}$, we can select either $q_i = 1$ or $q_i = 2$. Similar overlapping regions exist for $q_i = 0$ and $q_i = 1$, for $q_i = 0$ and $q_i = \bar{1}$, and also for $q_i = \bar{1}$ and $q_i = \bar{2}$. $\square$

In general, the ratio $k = \alpha/(\beta - 1)$ is a measure of the redundancy in the representation of the quotient digits. The larger this ratio, the larger the overlap regions are in the plot of r_i/D versus $\beta r_{i-1}/D$. For example, if we set $\alpha = \beta - 1 = 3$, then k equals 1, which corresponds to the maximum redundancy. In this case, the region for $q_i = 1$ is $0 \leq \frac{4r_{i-1}}{D} \leq 2$, and, for $q_i = 2$, it is $1 \leq \frac{4r_{i-1}}{D} \leq 3$. Thus, the overlapping region where either $q_i = 1$ or $q_i = 2$ can be

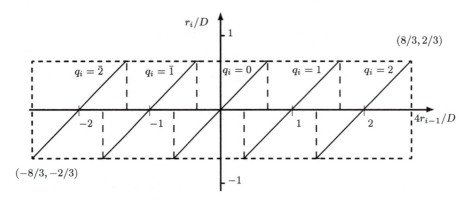

FIGURE 7.5 The quotient digits for $\beta = 4$ and $\alpha = 2$.

selected is $1 \leq \frac{4r_{i-1}}{D} \leq 2$, which is larger than the corresponding overlap region for $\alpha = 2$ (see Figure 7.5).

The implication of having an overlap region is that we have a choice of values, of both the partial remainder and the divisor, that will eventually separate the two adjacent regions corresponding to two consecutive values of the quotient digit q. The selected values of the partial remainder and divisor separating the adjacent regions of q will serve as comparison constants during the execution of the divide operation. We may therefore select these comparison constants so that they require as few digits as possible. Such a selection will reduce the execution time of the comparison step when determining the quotient digit.

Clearly, a larger overlap region (corresponding to a higher value of α) may allow us to select comparison constants with fewer digits. On the other hand, a higher value of α means having to produce more multiples of the form $\alpha \cdot D$, requiring extra hardware and/or time.

For a given α we need to determine the number of bits of the partial remainder and the divisor that must be examined in order to select the quotient digit. This is the most difficult step when developing a high-radix SRT algorithm. It can be accomplished numerically, analytically, or graphically. A combination of these techniques can also be employed.

To graphically determine the required number of bits of the partial remainder and the divisor we use the P-D or partial remainder versus divisor plot ([2, 9]), like the one shown in Figure 7.6. The purpose of the P-D plot is to indicate the regions in which given values of q_i may be selected. To determine the region for a given value q of the quotient digit, consider again the basic equation for the partial remainder, rewritten as

$$\beta r_{i-1} = r_i + q \cdot D. \tag{7.10}$$

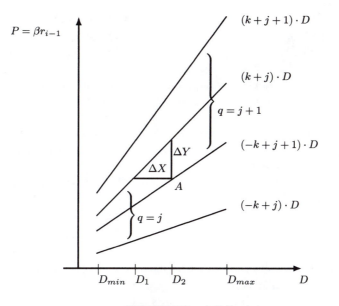

FIGURE 7.6   A P-D plot.

To simplify the notation from this point on, we denote the previous partial remainder βr_{i-1} by P, as shown in Figure 7.6. The maximum value of P for which the value q can be selected depends on the maximum allowed value for r_i and since

$$- k\, D \;\leq\; r_i \;\leq\; k\, D, \tag{7.11}$$

we obtain an upper limit for P, for which we may select the value q for the quotient digit:

$$P_{max} \;=\; (\,k \,+\, q\,)\cdot D \tag{7.12}$$

Similarly, the lower limit for P is

$$P_{min} \;=\; (\,-\,k \,+\, q\,)\cdot D. \tag{7.13}$$

Equations (7.12) and (7.13) for a specific value of q, say, $q = j$, are represented by two lines in Figure 7.6.

At each point in the region between these two lines we may select the value j for the quotient digit q. Due to the redundancy in representing the quotient, the regions for $q = j$ and $q = j + 1$ overlap. The overlapping region is between the upper line for $q = j$ and the lower line for $q = j + 1$, as shown in Figure 7.6. Note that Figure 7.6 includes only positive values of the divisor and partial remainder, and thus constitutes only one-quarter of the complete P-D plot. However, the complete P-D plot is symmetric about both axes, and, in most cases, it is sufficient to analyze one-quarter of it. Note also that only values of $|D|$ in the range $[D_{min}, D_{max}]$ are of interest. Examples for this range are $[0.5, 1)$ and $[1, 2)$. The latter is applicable when dividing floating-point numbers in the IEEE standard (see Chapter 4).

For the above-mentioned overlapping region, we have to determine the value of P, which will eventually separate the selection regions of $q = j$ and $q = j+1$. This value will serve as a comparison constant, and the number of bits required to represent it will determine the necessary precision when examining the partial remainder in order to select q. The line separating the regions of the partial remainder may be a straight horizontal line (one that is independent of D) or a stairstep function, partitioning the range of the divisor $[D_{min}, D_{max})$ into several intervals. We may have a single horizontal line $P = c$, where c is a constant, if and only if there is a value c satisfying

$$(k + j) \cdot D_{min} \geq c \geq (-k + j + 1) \cdot D_{max}. \tag{7.14}$$

This implies that the selection of q will be independent of D and will depend only on P. If this inequality is not satisfied, we must divide $[D_{min}, D_{max})$ into several smaller intervals. The "stepping" points determine the precision (i.e., the number of digits) at which we have to examine D, while the height of the steps determines the precision at which the partial remainder has to be examined. The maximum width of a step between D_1 and D_2, denoted by ΔX, is the horizontal distance between the two lines defining the overlap region (see Figure 7.6). The expression for this horizontal distance is, in general,

$$\Delta X = D_2 - D_1 = \frac{P}{-k + j + 1} - \frac{P}{k + j}$$

$$= P \cdot \frac{2k - 1}{j(j + 1) + k(1 - k)}. \tag{7.15}$$

The horizontal distance ΔX is minimal when j is maximal and P is minimal. The maximum value of $(j + 1)$ is α; hence $j = \alpha - 1$; P is minimal when $D_1 = D_{min}$. Thus,

$$\Delta X_{min} = D_{min} \cdot (k + \alpha - 1) \frac{2k - 1}{\alpha(\alpha - 1) + k(1 - k)}. \tag{7.16}$$

The maximum height ΔY of the step, or the vertical distance between the two lines, is

$$\Delta Y = (k + j)D - (-k + j + 1)D = (2k - 1) \cdot D. \tag{7.17}$$

This vertical distance is a minimum when $D = D_{min}$. Consequently, to determine the precision at which the partial remainder and divisor have to be examined, it is sufficient to consider the overlapping region between $q = \alpha$ and $q = \alpha - 1$ near D_{min}.

Let N_P denote the number of bits of the partial remainder P that have to be examined in order to determine the correct value q of the quotient digit. N_D is defined similarly for the divisor. The selection of the value q can be done by a look-up table implemented for example, in a PLA (programmable logic array) with $N_P + N_D$ inputs. Our objective is to minimize the size of the look-up table, which, in turn, will speed up the division process.

Let ϵ_P and ϵ_D denote the number of fractional bits within N_P and N_D, respectively. The precision at which the partial remainder is examined is thus $2^{-\epsilon_P}$, and similarly $2^{-\epsilon_D}$ is the precision of the "truncated" divisor. Clearly, these two must satisfy

$$2^{-\epsilon_D} \leq \Delta X_{min} \quad \text{and} \quad 2^{-\epsilon_P} \leq \Delta Y_{min}. \tag{7.18}$$

However, these inequalities provide only upper bounds for determining the precision at which the divisor and remainder should be examined, since the two extreme points of the interval ΔX (ΔY) may require a higher precision; i.e., more than ϵ_D (ϵ_P) fractional bits. To check whether the computed values of ϵ_D and ϵ_P are sufficient or whether a higher precision is required, we may use the P-D plot. This plot can also be used to decide on the value of q for each pair of values of P and D when truncated to the most significant N_P and N_D bits, respectively.

When making these decisions we must take into account the limited precision of the divisor and the partial remainder. As a result, each point with the coordinates (P, D) in the P-D plot represents all the partial remainder-divisor pairs with

$$P \leq \text{ partial remainder } \leq P + 2^{-\epsilon_P},$$

$$D \leq \text{ divisor } \leq D + 2^{-\epsilon_D}.$$

Therefore, we must select for the pair (P, D) a value of q which is legitimate for all the pairs in the above range.

Consider, in particular, point A in Figure 7.6. For a divisor that equals D_2 we may select $q = j + 1$, but for a divisor that equals $D_2 + 2^{-\epsilon_D}$ we must select $q = j$. Consequently, we should not select $q = j + 1$ for point A or for any other point in the overlap region whose horizontal distance from the line $(-k + j + 1)D$ is smaller than $2^{-\epsilon_D}$.

Example 7.7

The P-D plot for $\beta = 4$, $\alpha = 2$, and $D \in [0.5, 1)$ is shown in Figure 7.7. In this figure, the overlapping region for $q = 1$ and $q = 2$ lies between the lines

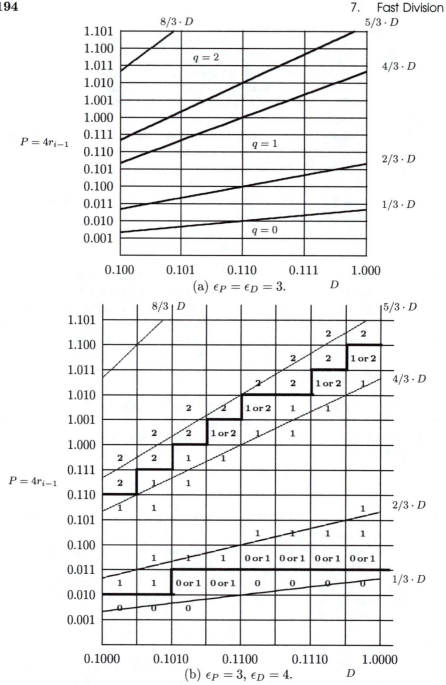

FIGURE 7.7 The P-D plot for $\beta = 4$, $\alpha = 2$, and $D \in (0.5, 1)$.

(1) $P = (k + \alpha - 1)D = 5/3 \cdot D$,   and

(2) $P = (-k + \alpha)D = 4/3 \cdot D$

We first check the possibility of a single horizontal line, as in Equation (7.14). $(k + 1)D_{min} = 5/3 \cdot 0.5$ and $(-k + 2)D_{max} = 4/3 \cdot 1$, and since $5/6 < 4/3$, a single line is impossible and we must have several divisor intervals. Next, we calculate the smallest horizontal and vertical distances:

$$\Delta X_{min} = D_{min} \cdot \frac{5}{3} \cdot \frac{3}{20} = \frac{5}{6} \cdot \frac{3}{20} = \frac{1}{8} = 2^{-3}, \quad \text{hence} \quad \epsilon_D \geq 3$$

$$\Delta Y_{min} = D_{min} \cdot 1/3 = 1/6, \quad \text{hence} \quad \epsilon_P \geq 3$$

The P-D plot in Figure 7.7(a) includes a grid with $\epsilon_P = \epsilon_D = 3$. A precise examination of the overlapping region between $q = 1$ and $q = 2$ in Figure 7.7(a), reveals that for the partial remainder-divisor pair $(0.110, .100)$ we cannot select a value of q which will be legitimate for all the points in the corresponding rectangle. Furthermore, it is apparent that by increasing ϵ_D to 4 the above problem is resolved. The corresponding grid is shown in Figure 7.7(b). We now have to decide on the value of q for each rectangle in this grid. Figure 7.7(b) shows all the possible selections of a value for q for all the pairs (P, D) within the overlapping regions. The heavy lines in the figure show one possible set of lines separating the regions for different values of q. Clearly, this is one out of many possible solutions allowing the designer to select a solution that will, for example, minimize the PLA which implements the look-up table for q. Such a PLA will have $N_D + N_P$ inputs where $N_D = 4$ and $N_P = 6$, since three more bits are needed for the integer part of the remainder and its sign $(-8/3 \leq P \leq 8/3)$. The number of inputs to this PLA can be reduced to $N_P + N_D - 1 = 9$ by taking advantage of the fact that the most significant bit of D is always 1 and can be therefore omitted.

Note that theoretically, for the overlapping region between $q = 1$ and $q = 0$ we could use a single horizontal line, since $2/3 \cdot 1/2 \leq 1/3$. This however, will require a high-precision comparison of the partial remainder, since $c = 1/3$ requires that all the fractional bits in the partial remainder will be compared. We therefore partition the divisor interval into two subintervals, as shown in Figure 7.7(b).   □

Example 7.8

Let $X = (0.00111111)_2 = 63/256$ and $D = (0.1001)_2 = 9/16$ as in Example 7.5. For this divisor, the comparison constants for the partial remainder are, according to Figure 7.7(b), 1/4 (0.010) and 7/8 (0.111).

```
r₀ = X          0  .0  0  1  1  1  1  1  1
4r₀          0  0  .1  1  1  1  1  1               ≥ 7/8  set q₁ = 2
Add −2D      1  0  .1  1  1  0
─────────────────────────────────────────────────────────────────
r₁           1  1  .1  1  0  1  1  1
4r₁          1  1  .0  1  1  1                     < −1/4  set q₂ = 1̄
Add D        0  0  .1  0  0  1
─────────────────────────────────────────────────────────────────
r₂           0  0  .0  0  0  0                     zero final remainder
```

The resulting quotient is $Q = 0.2\bar{1}_4 = 0.100\bar{1}_2 = 0.0111_2 = 7/16$.   □

The entries of the look-up table for the above algorithm can also be calculated numerically (instead of graphically) with the number of inputs to the look-up table determined by a trial-and-error numerical search. Suppose, for example, that we start with an initial guess of $\epsilon_D = 3$ and $\epsilon_P = 3$ and we attempt to calculate the value of q for $D = 0.100$ and $P = 0.110$. Since we truncated the divisor, we need to consider divisors from 0.100 to 0.101. Similarly, the partial remainder could have a value from 0.110 to 0.111. Thus, P/D could be as small as 0.110/0.101 or as large as 0.111/0.100. The first equals $0.110/0.101 = 1.2$ and, according to Figure 7.5, requires $q = 1$, while the second equals $0.111/0.100 = 1.75$, requiring $q = 2$. We therefore conclude that the above precision of P and D is insufficient. We must increase the number of bits of either the divisor or the partial remainder and try again. A simple program could be prepared to perform this numerical search and determine the value of q for each (P, D) pair [6].

To illustrate the lower precision of comparison needed for a higher value of α (i.e., a higher level of redundancy) consider the following example.

Example 7.9

For $\beta = 4$ and $\alpha = 3$, k is $\alpha/(\beta - 1) = 1$. The region for $q = 2$ is between the lines $P = (k + q)D = 3D$ and $P = (-k + q)D = D$, while the region for $q = 3$ is between the lines $P = 4D$ and $P = 2D$. Thus, the overlapping region for $q = 2$ and $q = 3$ is between the lines $P = 3D$ and $P = 2D$, as shown in Figure 7.8. For $D \in [1, 2)$, as in the IEEE floating-point standard, we have the following inequalities:

$$\Delta X_{min} = D_{min} \cdot 3 \cdot \frac{1}{6} = \frac{3}{6} = 2^{-1}, \quad \text{hence} \quad \epsilon_D \geq 1;$$

$$\Delta Y_{min} = D_{min} \cdot 1 = 1, \quad \text{hence} \quad \epsilon_P \geq 0.$$

To obtain the values of the comparison constants we have to examine the diagram in Figure 7.8. We conclude that $\epsilon_D = 1$ and $\epsilon_P = 0$ and therefore,

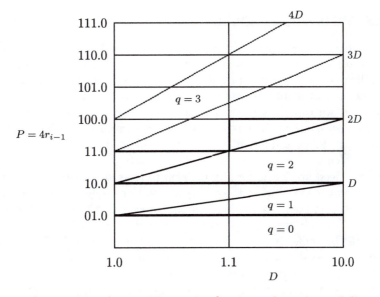

FIGURE 7.8   The P-D plot for $\beta = 4$, $\alpha = 3$, and $D \in (1,2)$.

$N_D = 2$ and $N_P = 4$, instead of $N_D = 4$ and $N_P = 6$ for $\alpha = 2$. A less complicated quotient selection logic is needed here, but the multiple $3D$ is required, which is costly to generate.   □

Example 7.10

Let $X = (01.0101)_2 = 21/16$ and $D = (01.1110)_2 = 15/8$. For this divisor, the partial remainder comparison constants are, according to Figure 7.8, 1, 2 and 4.

$r_0 = X$				0	1	.0	1	0	1	
$4r_0$		0	1	0	1	.0	1	0		≥ 4.0 set $q_1 = 3$
Add $-3D$ +	1	0	1	0	.0	1	1			
r_1		1	1	1	1	.1	0	1		
$4r_1$		1	1	1	0	.1	0	0		≥ -2.0 set $q_2 = \bar{1}$
Add D +	0	0	0	1	.1	1	1			
r_2		0	0	0	0	.0	1	1		final remainder $= 3/8 \cdot 2^{-4}$

The quotient is $Q = (0.3\bar{1})_4 = (0.110\bar{1})_2 = 11/16$. We verify the result of the divide operation through $X = Q \cdot D + R = 11/16 \cdot 15/8 + 3/128 = 168/128 = 21/16$.   □

7.3 SPEEDING UP THE DIVISION PROCESS

A major reason for the low speed of the division process is the fact that we have to complete the ith step before continuing to the $(i+1)$ step. The multiply process can be accelerated easily by generating several partial products simultaneously, since they are independent. In contrast, the steps in the division process are dependent, and we cannot start a new step before the current remainder is known and a new quotient digit is selected. Each step in the division consists of two substeps. First, a quotient digit is selected, and then the new partial remainder is calculated.

We can speed up the high-radix division process described in Section 7.2 in one of two ways. One way is to overlap the full-precision calculation of the partial remainder in step i with the selection process of the quotient digit in step $(i + 1)$. This overlapping is possible, since not all bits of the new partial remainder must be known in order to select the next quotient digit. Another way to speed up this process is to replace the carry-propagate add/subtract operation for calculating the new partial remainder by a carry-save operation.

In the first method, a truncated approximation of the new partial remainder is calculated in parallel to the full-precision calculation of the partial remainder. This approximation can be obtained at a high speed, enabling us to prepare for the new step (i.e., determine the quotient digit) even before the current step is completed.

Therefore, instead of first completing the calculation of the partial remainder r_{i-1} (with complete carry propagation) in step $(i-1)$, and then inputting the N_P most significant bits to the PLA to determine q_i in step i, we can use a small, fast adder that has as inputs the most significant bits of the previous partial remainder, denoted by $\widehat{\beta r_{i-2}}$, and the most significant bits of the corresponding multiple of the divisor, denoted by $\widehat{q_{i-1}D}$, as depicted in Figure 7.9.

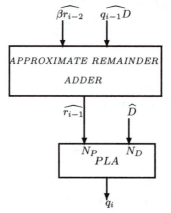

FIGURE 7.9   A quotient digit selection logic.

This approximate partial remainder (APR) adder produces an approximation of the required N_P most significant bits of the new partial remainder, denoted by $\widehat{r_{i-1}}$, before the full-precision add/subtract operation ($r_{i-1} = \beta r_{i-2} - q_{i-1}D$) is completed. This allows us to perform a look-ahead quotient digit selection, and we can select q_i in parallel with the full-precision calculation of the partial remainder r_{i-1}. Clearly, the size of this APR adder should be determined so that sufficiently accurate N_P bits will be generated. Since the uncertainty in the result of this adder is larger than the uncertainty in the truncated previous partial remainder βr_{i-2}, we may need additional input bits to the quotient digit look-up table.

Example 7.11

For $\beta = 4$, $\alpha = 2$, and $D \in [1, 2)$, the P-D plot is shown in Figure 7.10. An APR adder of the convenient size of eight bits is sufficient to generate the necessary inputs to the quotient selection PLA [17]. The horizontal lines in Figure 7.10 were determined to reduce the complexity of the PLA. Only three divisor bits are needed as inputs to the PLA, since the most significant bit of the divisor is always 1. As for the partial remainder, out of the outputs of the APR adder, five bits (including the sign bit) are sufficient in most cases. For a positive partial remainder, only in three cases (marked by a, b, and c in Figure 7.10) is an additional bit required. In case a, $D = 1.001$ and $P = 1.1$ and the single fractional bit of P is insufficient. The divisor can assume any value from 1.001 to 1.010. The partial remainder can have a value from 1.1 to 10.0. Thus, the range for P/D is from $1.1/1.010 = 1.2$ to $10.0/1.001 = 1.77$. The first requires $q = 1$, while the second requires $q = 2$ (see Figure 7.5). Adding a second fractional bit to P solves the problem, allowing us to select $q = 1$ for $P = 1.10$ and $q = 2$ for $P = 1.11$.

In cases b and c, the extra fractional bit of P is required, since the 8-bit APR adder may introduce an additional truncation error, further increasing the range of P/D. Consider for example, case b, where $D = 1.100$ and $P = 10.0$. If no APR adder is used, the range for P/D is from $10.0/1.101 = 1.23$ to $10.1/1.100 = 1.66$ and $q = 1$ can be selected. The 8-bit APR adder introduces an error of up to 2^{-6} in r_{i-1}, which increases to 2^{-4} after the multiplication by 4. This additional error increases the maximum value of P/D from 1.66 to 1.7, requiring q to be 2. An extra fractional bit of P solves this problem.

For a negative partial remainder represented in two's complement, there are six cases where 1 or even 2 additional output bits of the APR adder are required to guarantee the correct selection of the quotient digit [17].

□

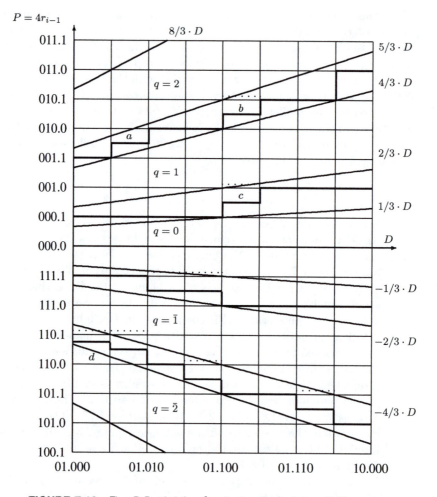

FIGURE 7.10   The P-D plot for $\beta = 4$, $\alpha = 2$, and $D \in (1,2)$.

In the scheme described above, the time needed for each step of the division is primarily determined by the time required to perform the add/subtract operation for calculating the new partial remainder, since the quotient digit was selected in the previous step. The second method for speeding up the division process avoids the time-consuming carry propagation when calculating the new partial remainder. Since a truncated partial remainder is sufficient for selecting the next quotient digit, there is no need to complete the calculation of the partial remainder at any intermediate step in the division. Thus, instead of using a carry-propagating adder to calculate the new partial remainder, we can use a carry-save adder and represent the partial remainder in a redundant form using

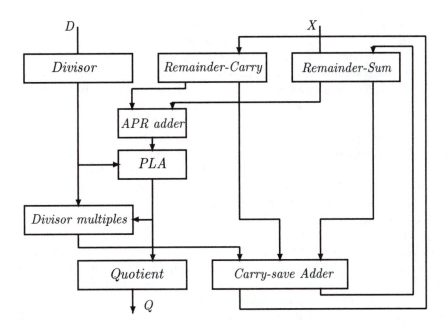

FIGURE 7.11 An SRT divider with redundant remainder.

two sequences of intermediate sum bits and carry bits. These should be stored in two separate registers as shown in Figure 7.11. Only the most significant sum and carry bits must be assimilated using the APR adder in order to generate an approximate partial remainder and allow the selection of the quotient digit. In this case, the calculation of the approximate partial remainder and the selection of the quotient digit are the most time-consuming operations. Thus, in each step of the division process a carry-save adder calculates the partial remainder, and the APR adder then accepts the most significant sum and carry bits of the partial remainder and generates the required inputs to the quotient selection PLA. As in the first method, the number of inputs to the PLA and its entries need to be calculated, taking into account the uncertainty in the sum and carry bits representing the truncated remainder.

Example 7.12

An algorithm for high-speed division with $\beta = 4$, $\alpha = 2$, and $D \in [1, 2)$ has been presented in [7]. The partial remainder is calculated in a carry-save manner and, consequently, two registers are needed to store the sum bits and the carry bits separately, resulting in a somewhat more complex design. An 8-bit APR adder is used to generate the most significant partial remainder bits that are needed as inputs to the quotient selection PLA.

The inputs to this APR adder are the eight most significant sum bits and carry bits in the redundant representation of the partial remainder. The outputs of the APR adder are then converted to a sign-magnitude representation and, as a result, only four bits of the approximate partial remainder are needed in most cases. Only in four cases is an additional bit required, yielding a very simple PLA.   □

Further speed up of the SRT division can be achieved by increasing the radix β of the algorithm to 8 or even higher. This reduces the number of steps to $\lceil n/3 \rceil$ or lower. Several such radix-8 SRT dividers have been implemented [8, 10]. The main disadvantage of the radix-8 (or higher) SRT algorithm is the high complexity of the quotient selection PLA which then becomes the most time-consuming unit of the divider in Figure 7.11.

One way to avoid the need for a very complex quotient selection PLA is to implement a radix-2^m SRT unit as a set of m overlapping radix-2 SRT stages. The radix-2 SRT requires a very simple quotient selection logic since q_i ($q_i \in \{-1,0,1\}$) is solely determined by the remainder and is independent of the divisor. We must however, overlap the quotient selections for the m bits so that all m quotient bits will be generated in one step of the process. Figure 7.12 depicts two overlapping radix-2 SRT stages which generate two quotient bits (q_i and q_{i+1}) in one step, implementing a radix-4 division.

Based on the most significant bits of the two remainder sequences (the sum and carry sequences), a value for q_i is generated using a Q_{sel} unit. In parallel, all three possible values of q_{i+1} are generated using three Q_{sel} units. These values correspond to the three possible intermediate remainders, namely, $2r_{i-1} - D$, $2r_{i-1}$ and $2r_{i-1} + D$. Note however that only the most significant bits of these three remainders have to be generated. Once q_i is known, the correct value of q_{i+1} is selected. This value is then used to select the correct multiple of the divisor to form the new remainder which will be stored in the two registers (for the sum and carry sequences). The overall delay of the radix-4 circuit in Figure 7.12 is determined by the delay of a Q_{sel}, the delay of two multiplexor units and the delay of the final CSA unit. This delay may be shorter than the delay of a radix-4 stage due to the higher complexity of the radix-4 quotient selection PLA [14].

Extending the above technique to radix-8 SRT division necessitates a more complex quotient selection circuit since three quotient digits (namely, q_i, q_{i+1} and q_{i+2}) must be generated in parallel. For generating q_{i+1} the speculative remainders $2r_{i-1} - D$, $2r_{i-1}$ and $2r_{i-1} + D$ have to be calculated. For generating q_{i+2} the speculative remainders $4r_{i-1} - 3D$, $4r_{i-1} - 2D$, $4r_{i-1} - D$, $4r_{i-1}$, $4r_{i-1} + D$, $4r_{i-1} + 2D$, and $4r_{i-1} + 3D$ must be calculated (again, only the most significant bits of these seven remainders). This implies that seven Q_{sel} units are required with multiplexors (controlled by q_i and q_{i+1}) to select the correct value

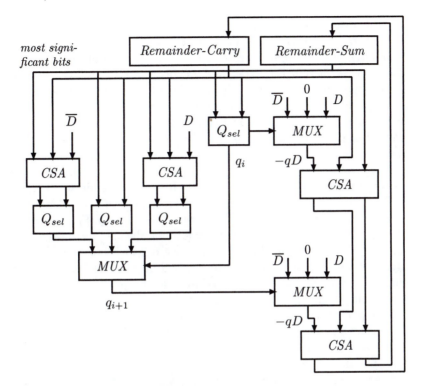

FIGURE 7.12 Two overlapping radix-2 SRT stages.

of q_{i+2} [14]. Extending the above to four overlapping radix-2 stages to obtain a radix-16 divider will result in an increase in the total number of Q_{sel} units from 11 to 26 making this approach more costly. Another alternative for a radix-16 divider is to use two overlapping radix-4 SRT stages [18].

7.4 ARRAY DIVIDERS

All algorithms for division can be implemented using an array of cells where each step of the algorithm is executed by a separate row of cells. Thus, n rows of cells with n cells per row are required to implement a radix-2 division algorithm. If the restoring algorithm is selected for implementation, then the difference in each row between the previous partial remainder and the divisor is formed and the quotient bit is generated according to the sign of this difference. There is no need, however, to restore the partial remainder if the quotient bit is determined to be 0. Instead, according to the generated quotient bit, either the previous partial remainder or the difference (which constitutes the new partial remainder) is transferred to the next row. If a ripple-carry scheme is employed in every row, then it takes n steps to propagate the carry in a single row and, since there are n rows, the total execution time is of the order of n^2.

Similarly, we can implement a nonrestoring division array. It has about the same speed as the restoring array, and its only advantage is its ability to handle negative operands in a simple way. On the other hand, the final remainder may be incorrect, having a sign opposite to that of the dividend. An example of such an array divider is shown in Figure 7.13, where $x_0.x_1 \cdots x_6$ is the dividend, $d_0.d_1d_2d_3$ is the divisor, $q_0.q_1q_2q_3$ is the quotient, and $r_0.r_1r_2r_3$ is the final remainder [3].

The operation (addition or subtraction) to be performed in a given row is controlled by the signal T (see Figure 7.13(a)). If $T = 0$ an addition is performed, while if $T = 1$, a subtraction is executed by adding the two's complement of the divisor, which is assumed (in the implementation depicted in Figure 7.13) to be positive (i.e., $d_0 = 0$). The latter is done by forming the one's complement of the divisor and forcing a carry of 1 into the rightmost cell by connecting it to T. The generated quotient and remainder are represented in two's complement, but the final remainder is not always correct. To show that the quotient generated by the array divider in Figure 7.13 is correct, note that in each step of the nonrestoring division, the partial remainder and the multiple of the divisor ($\pm D$) always have opposite signs. Therefore, the carry-out from the leftmost cell equals 1 if the sign of the new partial product is 0, and vice versa. Hence, $c_{out} = 1$ implies that the operation in the next row should be subtraction ($T = 1$), since the divisor is assumed to be positive. Similarly, $c_{out} = 0$ generates $T = 0$, so an add operation should be performed in the next row. The values $c_{out} = 1$ and $c_{out} = 0$ for row i correspond to $u_{i+1} = 1$ and $u_{i+1} = \bar{1}$, respectively, where u_{i+1} is the $(i+1)$th quotient bit. This is identical to the relationship between p and q in the algorithm for converting the representation of the quotient that uses the digit set $\{\bar{1}, 1\}$ to the equivalent two's complement representation, as described in Section 3.3.

In the two previously presented array dividers a complete add/subtract operation with carry-propagation is performed by each row in the array. In the nonrestoring division, only the sign bit of the partial remainder is needed to select the quotient bit. This sign bit can be generated by using a fast carry-look-ahead circuitry, while the other bits of the partial remainder can be generated using carry-save adders. Each cell generates a P_i and G_i output (carry-propagate and carry-generate, respectively, as in a carry-look-ahead adder) in addition to the ordinary sum and carry outputs. The P_i and G_i outputs of all cells in the same row are connected to a carry-look-ahead circuit, which generates the quotient bit. The execution time of such an array divider is of the order of $n \log n$, compared to n^2 for the previous two array dividers [3].

In a similar way, we can implement a high-radix division array with carry-save addition. Here, a small carry-look-ahead adder is used to determine the

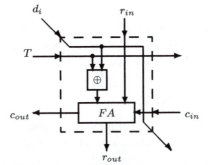

(a) Controlled add/subtract (CAS) cell.

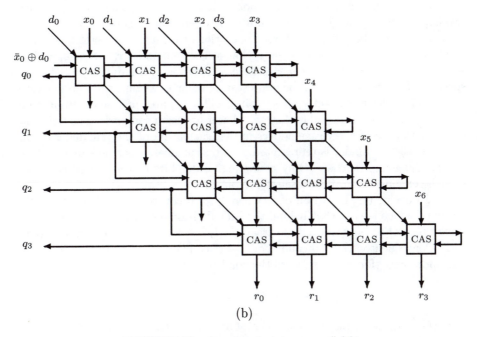

(b)

FIGURE 7.13 A nonrestoring array divider.

correct most significant bits of the partial remainder in order to select the quotient digit. Due to the similarity between the basic cells in array multipliers and array dividers, a controlled array multiplier/divider can be designed. Several such circuits have been described; e.g., [1].

7.5 FAST SQUARE ROOT EXTRACTION

As pointed out in Section 3.4, the similarities between square root extraction and the divide operation allow the adaptation of almost all the algorithms that have been developed for division to the calculation of the square root, with only some minor modifications. Consequently, small extensions to the hardware designed for a division unit enable the calculation of the square root (e.g., [7], [17], [21]).

The nonrestoring algorithm for square root extraction that has been presented in Section 3.4 allows the use of the digits 1 and $\bar{1}$ for q_i, where $Q = 0.q_1, \cdots, q_m$ is the calculated square root. Allowing q_i to assume the value 0 has two important advantages. First, a shift-only operation is required when $q_i = 0$, reducing the number of add/subtract operations that must be performed. Second, having a overlap between the region of the remainder r_i where $q_i = 1$ is selected and the region where $q_i = 0$ is selected leads to a reduced precision inspection of the remainder. In the nonrestoring scheme, we must identify the case of $r_i \geq 0$ in order to correctly set the bit q_i. This requires precise determination of the sign bit of r_i. If $q_i = 0$ is allowed, a lower precision comparison is sufficient, enabling the use of carry-save adders in the calculation of the remainder r_i. In this case, the remainder is represented as two sequences, a partial sum sequence and a sequence of carries. Only a few high-order bits of these two sequences must be examined in order to select the bit q_i.

To decide on the region of r_i where $q_i = 0$ can be selected, we basically follow the same idea as that behind the SRT division algorithm. We restrict the square root Q to be a normalized fraction, $1/2 \leq Q < 1$, with q_1 always equal to 1. Consequently, the radicand should satisfy $1/4 \leq X < 1$. In this case, the remainder r_{i-1} (for $i \geq 2$) satisfies the condition [12]

$$-2(Q_{i-1} - 2^{-i}) \leq r_{i-1} \leq (Q_{i-1} + 2^{-i}),$$

where Q_{i-1} is the partially calculated root at step $(i-1)$, i.e., $Q_{i-1} = 0.q_1 q_2 \cdot q_{i-1}$. In step $i \geq 2$, we may therefore set q_i equal to 0, which results in $r_i = 2r_{i-1}$ whenever r_{i-1} is in the range $[-(Q_{i-1} - 2^{-i-1}), (Q_{i-1} + 2^{-i-1})]$. Hence, a possible selection rule for q_i is

$$q_i = \begin{cases} 1 & \text{if} \quad r_{i-1} \geq (Q_{i-1} + 2^{-i-1}) \\ 0 & \text{if} \quad -(Q_{i-1} - 2^{-i-1}) \leq r_{i-1} \leq (Q_{i-1} + 2^{-i-1}) \\ \bar{1} & \text{if} \quad r_{i-1} \leq -(Q_{i-1} - 2^{-i-1}). \end{cases} \qquad (7.19)$$

Since $(Q_{i-1}+2^{-i-1})$ and $(Q_{i-1}-2^{-i-1})$ are in the range $[1/2, 1]$, we may replace (7.19) by the following selection rule, which avoids a high-precision comparison:

$$q_i = \begin{cases} 1 & \text{if} \quad 1/2 \leq 2r_{i-1} \leq 2 \\ 0 & \text{if} \quad -1/2 \leq 2r_{i-1} < 1/2 \\ \bar{1} & \text{if} \quad -2 \leq 2r_{i-1} < -1/2. \end{cases} \qquad (7.20)$$

This selection rule is similar to the SRT rule in Equation (7.4).

Example 7.13

Let $X = 0.0111101_2 = 61/128$:

$r_0 = X$			0.0111101	
$2r_0$			0.1111010	set $q_1 = 1,\ Q_1 = 0.1$
$-(0 + 2^{-1})$	$-$		0.1000000	
r_1			0.0111010	
$2r_1$			0.1110100	set $q_2 = 1,\ Q_2 = 0.11$
$-(2Q_1 + 2^{-2})$	$-$	0	1.0100000	
r_2			1.1010100	
$2r_2$		1	1.0101000	set $q_3 = \bar{1},\ Q_3 = 0.101$
$+(2Q_2 - 2^{-3})$	$+$	0	1.0110000	
r_3		0	0.1011000	
$2r_3$		0	1.0110000	set $q_4 = 1,\ Q_4 = 0.1011$
$-(2Q_3 + 2^{-4})$	$-$	0	1.0101000	
r_4		0	0.0001000	
$2r_4$		0	0.0010000	set $q_5 = 0,\ Q_5 = 0.10110$
r_5		0	0.0010000	
$2r_5$		0	0.0100000	set $q_6 = 0,\ Q_6 = 0.101100$
r_6		0	0.0100000	
$2r_6$		0	0.1000000	set $q_7 = 1,\ Q_7 = 0.1011001$
$-(2Q_6 + 2^{-7})$	$-$	0	1.0110001	
r_7		1	0.0001111	

The square root is $Q = 0.1011001 = 89/128$. The final remainder is $2^{-7}r_7 = -113/2^{14} = X - Q^2 = (7808 - 7921)/2^{14}$.   □

The high-radix algorithms for division can also be modified to calculate the square root. The general equation for computing the new remainder is

$$r_i = \beta r_{i-1} - q_i \cdot (2Q_{i-1} + q_i\beta^{-i}) \qquad (7.21)$$

where β is the radix and the digit set for q_i is $\{\bar{\alpha}, \overline{\alpha - 1}, \cdots, \bar{1}, 0, 1, \cdots, \alpha\}$, as it is for division.

For example, when $\beta = 4$, the digit set $\{\bar{2}, \bar{1}, 0, 1, 2\}$ is preferable, since it eliminates the need to generate the multiple $3Q_{i-1}$. The generation of the root multiple $q_i \cdot (2Q_{i-1} + q_i 4^{-i})$ makes the square root extraction somewhat more complex than the equivalent division. However, a careful examination of the required multiples shows that these can be easily calculated. For the positive values of q_i, namely, 1 and 2, we have to subtract the sequences $Q001_2$ and $Q010_2$, respectively. When $q_i = \bar{1}$, we have to add the sequence $Q00\bar{1}_2$, which has the same value as $(Q - 1)111$ [7]. Similarly, when q_i is $\bar{2}$, we must add the sequence $Q0\bar{1}0_2$, which is equivalent to $(Q - 1)110_2$. Thus, having two registers with the values Q and $Q - 1$, updated at every step, greatly simplifies the execution of the square root algorithm [7].

Since only a low-precision comparison of the remainder is needed in order to select the quotient digit, we may perform the add/subtract operation in Equation (7.21) in a carry-save manner, and use a small carry-propagate adder to calculate the most significant bits of r_i. Those will then provide some of the inputs to a PLA for selecting the square root digit q_i, similarly to division. The other inputs to the PLA are the most significant bits of the root multiple. Several rules for selecting q_i have been proposed ([4], [7], and [21]). In these rules, the intervals of the remainder determine the size of the carry-propagating adder (between 7 and 9 bits for the base-4 algorithm with the digits $\{\bar{2}, \bar{1}, 0, 1, 2\}$) and the exact PLA entries. The selected digit q_i depends on the truncated remainder and the truncated root multiple. In the first step, however, no estimated root is available. Consequently, a separate PLA for predicting the first few bits of the root may be necessary.

Example 7.14

The radix-4 divider (for double-precision floating-point numbers in the IEEE standard) reported in [7] is capable of calculating the square root as well. The P-D plot in Figure 7.10 was also used for square root extraction. Therefore, the same PLA (with 19 product terms) is used for predicting the next quotient digit and the next root digit. A separate PLA (with 28 product terms) was added to the arithmetic unit in order to provide the five most significant bits of the root. The inputs to this PLA are the six most significant bits of the significand and the least significant bit of the exponent. The latter indicates whether the exponent is odd or even. This is necessary in order to find the square root, according to the following equation:

$$\sqrt{1.f \cdot 2^{E-1023}} = \begin{cases} \sqrt{0.1f} \cdot 2^{(E-1)/2 - 1023} & \text{if } E \text{ is odd} \\ \sqrt{0.01f} \cdot 2^{E/2 - 1 - 1023} & \text{if } E \text{ is even} \end{cases} \tag{7.22}$$

Note that the resulting radicand ($\sqrt{0.1f}$ or $\sqrt{0.01f}$) is in the range $[1/4, 1]$ yielding a square root in the required range, $[1/2, 1]$. $\square$

7.6 EXERCISES

7.1. Show that, for a divisor $D = 3/4$, the SRT algorithm will always generate the minimal SD representation of the quotient.

7.2. Given a dividend $X = 0.1001$ and a divisor $D = 0.1010$, perform the divide operation using **(a)** the standard binary SRT algorithm, **(b)** the modified binary SRT algorithm allowing the use of the multiples $D/2$ and $2D$, and **(c)** the modified binary SRT algorithm with five different comparison constants.

7.3. Check whether the divisor is within the favorable region (requiring fewer add/subtract operations) for all five comparison constants $K_1, \cdots K_5$.

7.4. Analyze the possibility of using the first two bits of the divisor as a comparison constant K to speed up the SRT algorithm for a divisor in the range $[0.5,1]$. For example, for $D = 0.1101$, the most significant bits of the partial remainder will be compared to $K = 0.11$.
(**a**) Find whether the number of add/subtract operations in the suggested method will be smaller than, equal to, or greater than the number of these operations in the original SRT method (with $K = 0.1$) in the following three cases:
(i) $D = 0.1001$ and $X = 0.10000111$
(ii) $D = 0.1101$ and $X = 0.01011011$
(iii) $D = 0.1111$ and $X = 0.01001011$
(**b**) Would you recommend the use of the above algorithm?

7.5. Derive a variation of the binary SRT division algorithm that uses several comparison constants similar to the one described in Section 7.1. Allow the comparison constants to have at most three fractional bits. The divisor domain $[0.5, 1)$ should be partitioned into subregions so that only the three most significant fractional bits of the divisor will be needed to determine the selected comparison constant.

7.6. Show that, for $\alpha = \beta - 1$, the high-radix SRT algorithm includes as a special case the high-radix nonrestoring division algorithm.

7.7. The following algorithm can be used to convert a quotient $q_1 q_2 \cdots q_n$, given in SD representation with the radix r and the allowed digit set $\{\overline{(r-1)}, \cdots \overline{1}, 1, \cdots (r-1)\}$, to its equivalent representation $y_0 y_1 y_2 \cdots y_n$ in radix complement:
Step 1: If $q_1 < 0$ then set $y_0 = r - 1$ and $y_1 = r + q_1$
            else set $y_0 = 0$ and $y_1 = q_1$
Step 2: Set $j = 1$.
Step 3: If $q_{j+1} < 0$ then set $y_j = y_j - 1$ and $y_{j+1} = r + q_{j+1}$
            else set $y_{j+1} = q_{j+1}$
Step 4: If $j = n - 1$ then stop
            else set $j = j + 1$ and goto step 3
(**a**) Use the above algorithm to convert the SD binary number $\overline{1}11\overline{1}$ to two's complement representation.
(**b**) Compare this algorithm to the one described in Section 3.3.
(**c**) Can the subtraction in step 3 generate a borrow that may propagate to higher-order bits?

7.8. Show the P-D plot for SRT division with $\beta = 4$, $\alpha = 3$ and $D \in [0.5, 1)$. Determine the regions for the different values of the quotient digit and the precision of the partial remainder and the divisor that is needed.

7.9. Show the second part of the P-D plot depicted in Figure 7.10 for a negative partial remainder in two's complement.

7.10. Redraw the P-D plot shown in Figure 7.10, allowing the required precision of the divisor and of the partial remainder to be 2^{-2} in most cases. Indicate the few cases where a higher precision is required.

7.11. Verify that the carry-out bits of the leftmost CASs in Figure 7.13 generate the correct quotient bits for $X = 0.011111$ and $D = 0.110$. Is the generated final remainder correct?

7.12. Can the array divider in Figure 7.13 be pipelined? What would the rate of such a pipeline be?

7.13. Show the block diagram of a binary nonrestoring array divider (with the same number of bits for the dividend and divisor, as in Figure 7.13) that generates the sign bit of the partial remainder using a carry-look-ahead circuit, while generating all other bits using carry-save addition. Design the basic cell(s) of the divider. Compare the execution time and implementation complexity of your array divider to those of the divider in Figure 7.13.

7.14. Show that exactly 11 radix-2 Q_{sel} units are required when implementing a radix-8 divider using three overlapping radix-2 SRT stages. Estimate the delay of these three overlapping stages.

7.7 REFERENCES

[1] D. P. AGRAWAL , "High-speed arithmetic arrays," *IEEE Trans. on Computers, C-28* (March 1979), 215-224.

[2] D. E. ATKINS, "Higher-radix division using estimates of the divisor and partial remainders," *IEEE Trans. on Computers, C-17* (Oct. 1968), 925-934.

[3] M. CAPPA and V. C. HAMACHER, "An augmented iterative array for high-speed binary division," *IEEE Trans. on Computers, C-22* (Feb. 1973), 172-175.

[4] L. CIMINIERA and P. MONTUSCHI, "Higher radix square rooting," *IEEE Trans. on Computers, 39* (Oct. 1990), 1220-1231.

[5] M. D. ERCEGOVAC and T. LANG, *Division and square root: Digit-recurrence algorithms and implementations*, Kluwer Academic Publishers, Norwell, 1994.

[6] D. GOLDBERG, "Computer arithmetic," in *Computer architecture: A quantitative approach*, D. A. Patterson and J. L. Hennessy, Morgan Kaufmann, San Mateo, CA, 1990.

[7] J. FANDRIANTO, "Algorithm for high speed shared radix 4 division and radix 4 square root," *Proc. of 8th Symp. on Computer Arithmetic* (1987), 73-79.

[8] J. FANDRIANTO, "Algorithm for high speed shared radix 8 division and radix 8 square root," *Proc. of 9th Symp. on Computer Arithmetic* (1989), 68-75.

[9] C. V. FREIMAN, "Statistical analysis of certain binary division algorithms," *Proc. of IRE, 49* (Jan. 1961), 91-103.

[10] R. F. HOBSON and M. W. FRASER, "An efficient maximum-redundancy radix-8 SRT division and square-root method," *IEEE Journal of Solid-State Circuits, 30* (Jan. 1995), 29-38.

[11] O. L. MACSORLEY, "High-speed arithmetic in binary computers," *Proc. of IRE, 49* (Jan. 1961), 67-91.

[12] S. MAJERSKI, "Square rooting algorithms for high-speed digital circuits," *IEEE Trans. on Computers, C-34* (Aug. 1985), 724-733.

[13] G. METZE, "A class of binary divisions yielding minimally represented quotients," *IRE Trans., EC-11* (Dec. 1962), 761-764.

[14] J.A. PRABHU and G.B. ZYNER, "167 MHz radix-8 divide and square root using overlapped radix-2 stages," *Proc. of the 12th Symp. on Computer Arithmetic* (July 1995), 155-162.

[15] J. E. ROBERTSON, "A new class of digital division methods," *IRE Trans. on Electronic Computers, EC-7* (Sept. 1958), 218-222.

[16] J. E. ROBERTSON, "The correspondence between methods of digital division and multiplier recording procedures," *IEEE Trans. on Computers, C-19* (Aug. 1970), 692-701.

[17] G. S. TAYLOR, "Compatible hardware for division and square root," *Proc. of 5th Symp. on Computer Arithmetic* (1981), 127-134.

[18] G. S. TAYLOR, "Radix 16 SRT dividers with overlapped quotient selection stages," *Proc. of 7th Symp. on Computer Arithmetic* (June 1985), 64-71.

[19] K. D. TOCHER, "Techniques of multiplication and division for automatic binary computers," *Quart. J. Mech. Appl. Math., 11*, Pt. 3 (1958), 364-384.

[20] J. B. WILSON and R. S. LEDLEY, "An algorithm for rapid binary division," *IRE Trans. on Electronic Computers, EC-10* (1961), 662-670.

[21] J. H. P. ZURAWSKI and J. B. GOSLING, "Design of a high-speed square root multiply and divide Unit," *IEEE Trans. on Computers, C-36* (Jan. 1987), 13-23.

8

DIVISION THROUGH
MULTIPLICATION

The number of steps in the previously described division methods was linearly proportional to the number of bits, n, while in the division-by-convergence schemes, which will be described in this chapter, the number of steps is proportional to $\log_2 n$. However, the basic operation in the division-by-convergence schemes is not an add/subtract operation but the usually slower multiply operation. Hence, a fast parallel multiplier is necessary to successfully implement these schemes.

8.1 DIVISION BY CONVERGENCE

Let the divisor D and the dividend N be considered the denominator and numerator, respectively, of the quotient Q. Thus, $Q = N/D$. This holds if we multiply both numerator and denominator by the same factor R_0, or even by m factors $R_0, R_1, \cdots, R_{m-1}$. If the factors R_i are selected so that the denominator converges to 1, the numerator will converge to Q:

$$Q = \frac{N}{D} = \frac{N \cdot R_0 R_1 \cdots R_{m-1}}{D \cdot R_0 R_1 \cdots R_{m-1}} \rightarrow \frac{Q}{1} \qquad (8.1)$$

In Equation (8.1) only the quotient is calculated, and a separate computation is necessary if the remainder is needed. Therefore, this division scheme is more suitable for floating-point computations.

The essential step in this method is the selection of the factors to ensure the convergence of the denominator to 1. This selection is based on the following observation: Let the divisor be a normalized binary fraction $0.1xxxx$ (where each x is either 0 or 1). Therefore, $1/2 \le D < 1$ and $D = 1 - y$, where $y \le \frac{1}{2}$. If we select R_0 to be $1 + y$, then the new denominator is

$$D_1 = D \cdot R_0 = (1 - y) \cdot (1 + y) = 1 - y^2, \tag{8.2}$$

and since $y^2 \le \frac{1}{4}$, the new denominator D_1 satisfies $D_1 \ge \frac{3}{4}$, and is therefore closer to 1 than D. In binary notation, the new denominator has the form $D_1 = 0.11xxxx$.

In step 2 we select $R_1 = 1 + y^2$ obtaining

$$D_2 = D_1 \cdot R_1 = (1 - y^2) \cdot (1 + y^2) = 1 - y^4. \tag{8.3}$$

Now, $y^4 \le \frac{1}{16}$ and, therefore, D_2 takes on the form $0.1111xxxx$, and is closer to 1 than D_1. In general, in the $(i + 1)$th step the divisor D_i has the form $D_i = 1 - y_i$, where $y_i = y^{2^i}$, and since $y \le 2^{-1}$, D_i has at least 2^i leading 1's. The next multiplying factor is $R_i = 1 + y_i$ and, as a result, $D_{i+1} = D_i R_i$ will have 2^{i+1} leading 1's.

To show formally that the denominator converges to 1, note that

$$(1 - y) \cdot [(1 + y)(1 + y^2)(1 + y^4) \cdots] = (1 + y) \cdot [(1 - y)(1 + y^2)(1 + y^4) \cdots]. \tag{8.4}$$

The term within the brackets on the right-hand side of Equation (8.4) is the series expansion of $1/(1 + y)$ for $0 \le y \le \frac{1}{2}$, hence

$$\lim_{i \to \infty} D_i = (1 + y) \cdot \frac{1}{1 + y} = 1. \tag{8.5}$$

The process of multiplying the numerator and denominator by R_i is repeated until D_i converges to 1, or, more precisely, to $0.11 \cdots 1$ (which equals $1 - ulp$). The number of leading 1's in D_i is doubled at each step, and therefore the number of iterations is $m = \lceil \log_2 n \rceil$, and we say that the convergence is *quadratic*. The multiplying factor in each step has to be generated and then multiplied by the numerator and denominator. The multiplying factor R_i has to be obtained from D_i. Fortunately, the relation between these two is simple. R_i equals $2 - D_i$; i.e., R_i is the two's complement of the fraction D_i. Thus, each step consists of the two multiplications

$$D_{i+1} = D_i \cdot R_i \quad \text{and} \quad N_{i+1} = N_i \cdot R_i \tag{8.6}$$

and a two's complement operation,

$$R_{i+1} = 2 - D_{i+1}. \tag{8.7}$$

Example 8.1

For the 15-bit numbers $N = 0.011\,010\,000\,000\,000 = 0.40625_{10}$ and $D = 0.110\,000\,000\,000\,000 = 0.75_{10}$ the quotient is calculated as follows:

The first multiplying factor is $R_0 = 2 - D = 1.010\,000\,000\,000\,000$,and as a result, the new numerator and denominator are $N_1 = N \cdot R_0 = 0.100\,000\,100\,000\,000$ and $D_1 = D \cdot R_0 = 0.111\,100\,000\,000\,000$. The second multiplying factor is $R_1 = 2 - D_1 = 1.000\,100\,000\,000\,000$,and the next numerator and denominator are $N_2 = N_1 \cdot R_1 = 0.100\,010\,100\,010\,000$ and $D_2 = D_1 \cdot R_1 = 0.111\,111\,110\,000\,000$. Note that the number of leading 1's in the denominator has doubled from 4 to 8. The third multiplying factor is $R_2 = 2 - D_2 = 1.000\,000\,010\,000\,000$, and then $N_3 = N_2 \cdot R_2 = 0.100\,010\,101\,010\,101$ and $D_3 = D_2 \cdot R_2 = 0.111\,111\,111\,111\,111$. Convergence has been achieved ($D_3 = 1 - ulp$) in three steps and the resulting quotient is $Q = N_3 = 0.54165_{10}$. The exact result is the infinite fraction 0.54166_{10}. $\qquad\square$

The total number of steps (order of $\log_2 n$) is smaller than that required by the algorithms based on add/subtract operations, for which the number of steps is linear in n. However, each step involves two multiplications, which are more time-consuming than add/subtract operations. Thus, there is a need to further reduce the number of steps. One way to achieve this is to speed up the first few steps, where the convergence is very slow. After step 1, only two leading 1's are guaranteed, and after step 2 is completed only four 1's are guaranteed. A speed up of the first steps can substantially reduce the required number of multiplications. Instead of selecting the first multiplier to be $R_0 = 1 + y$, we may use a look-up table that provides a multiplier that ensures a denominator D_1 with k leading 1's, where k is at least 3. The next denominator will then have $2k$ leading 1's, and so on. The size of this table, which can be stored in a ROM, increases exponentially with k, the desired number of leading 1's. The number k therefore has to be determined so that the required size of the table is still reasonable.

Example 8.2

The division scheme described above was first implemented in the IBM 360/91 for floating-point division [1]. The long format of floating-point numbers in the IBM system consists of 64 bits partitioned as follows:

S	7 bits - biased exponent	56 bits - unsigned fractional significand

The operands for the division are therefore 56-bit long fractions. If the first multiplying factor is $R_0 = 1 + y$, then $\lceil log_2 56 \rceil = 6$ steps are needed, requiring 12 multiplications of 56 bits each. Actually, only 11 multiplications are needed, since there is no need to calculate D_5 in the last step; we know that it equals 1 and there is no need to calculate another multiplying factor. To reduce the number of multiplications we can use a look-up table for the first factor R_0 so that D_1 will have at least $k = 7$ leading 1's, since it leads to the sequence

$$1 \;\rightarrow\; 7 \;\rightarrow\; 14 \;\rightarrow\; 28 \;\rightarrow\; 56$$

of leading 1's. Consequently, we need only four steps, requiring seven multiplications. For this k, it has been determined that 7 bits of D are needed and 10 bits of the factor R_1 have to be stored at each location, requiring a ROM of size 128×10. The contents of the row that corresponds to a given denominator $D = 1 - y$ is the approximated value of $(1+y)(1+y^2)(1+y^4)$ truncated to 10 bits. At its highest precision, such a multiplier, R_0, would guarantee eight leading 1's. It should be clear that no error is introduced by using a multiplier out of a table, since it is used to multiply both numerator and denominator, and the previous convergence scheme is initiated at this point. □

Another way to further reduce the execution time of the division algorithm is to speed up the multiplications by using shorter multipliers, since the multiplication time increases linearly with the number of multiplier bits. Instead of using a multiplier of length n bits for all multiplications, we can use a truncated multiplier for some of the products. Using a truncated multiplier will not introduce errors into the division process, since the numerator and denominator are multiplied by the same factor. Clearly, we cannot use a truncated multiplier for the last product because a high accuracy is needed at this point.

At step $(i+1)$ we want to generate D_{i+1} with a $(a \geq 2^{i+1})$ leading 1's by multiplying the denominator D_i, which has $a/2$ leading 1's, by R_i. Originally, $R_i = 2 - D_i = 1 + y_i$ (where $D_i = 1 - y_i$). Instead of using R_i, we use a truncated multiplier R_{i_T} by forming the two's complement of only the first a bits of D_i, which constitute the truncated denominator D_{i_T}. In other words, $R_{i_T} = 2 - D_{i_T}$. If we use the notation $R_{i_T} = 1 + y_T$, then the error in the truncated multiplier is $\alpha = y_T - y_i$. This "error" is always positive and satisfies the inequality $0 \leq \alpha < 2^{-a}$. When multiplying the truncated multiplier by the untruncated denominator we obtain

$$D_{i+1} = D_i \cdot R_{i_T} = (1 - y_i) \cdot (1 + y_T) = 1 + y_T - y_i - y_i y_T. \qquad (8.8)$$

Substituting $y_T = y_i + \alpha$ yields

$$D_{i+1} = 1 + \alpha - y_i(y_i + \alpha) = 1 - y_i^2 + \alpha(1 - y_i). \qquad (8.9)$$

The "error" in D_{i+1} is $\alpha(1 - y_i)$, which satisfies

$$0 \le \alpha(1 - y_i) \le \alpha < 2^{-a},$$

and we can therefore still expect to obtain a leading 1's in D_{i+1}. The only difference is that now D_{i+1} may converge toward 1 from either below (i.e., $D_{i+1} = 0.11 \cdots 1xxxx$) or above (i.e., $D_{i+1} = 1.00 \cdots 0xxxx$), since the "error" is always positive (see exercise 1).

In each of the truncated multiplication factors, the first half of the bits are identical, either all 0's or all 1's. If the multiplier R_i is recoded using *SD* representation, then these leading 0's or 1's will not generate nonzero partial products and the execution time of the multiplications will be further reduced.

Example 8.3

The fast multiplier in the floating-point arithmetic unit in the IBM 360/91 computer [1] receives operands of length 56 bits each and uses the algorithm outlined in Table 6.5 to generate partial products of the form 0, $\pm 2A$, and $\pm 4A$, where A is the multiplicand. The resulting 28 partial products require 26 carry-save adders. To reduce the amount of hardware needed, a smaller carry-save addition tree for eight operands was designed. This tree is capable of accepting six new partial products and adding them to the two previous intermediate results (intermediate sum and carry), which are connected through two feedback paths as shown in Figure 5.31. This tree must be used five times in order to accumulate all 28 partial products. The carry-save tree was designed as a pipeline to allow overlapping between consecutive sets of six partial products so that the accumulation of all 28 partial products would take only six clock cycles. This overlapping is possible among the sets of partial products that correspond to the same multiply operation. Complete overlapping between two different multiply operations is also achievable as long as the number of generated partial products is less than or equal to six. In such cases, the carry-save tree is passed only once per multiply operation, and there is no need to use the feedback connections. Consequently, limiting the number of partial products to six or less can significantly speed up the execution of the several consecutive multiplications needed in the division-by-convergence algorithm.

To achieve the sequence of denominators D_i with 7, 14, 28, and 56 leading ones (or zeros), we need multipliers R_i with 10, 14, 28, and 56 bits, respectively. The first multiplier is read out of the ROM and generates only five partial products. The other three multipliers contain 7, 14, and 28 leading zeros (or ones) which can be skipped, and there is no need to generate any partial products for them. We only need to identify the first and last bits of such a group of identical bits. Therefore, the second multiplier, of length 14 bits, generates only the five partial products

$$0. \underbrace{11}\ 11\ 11\ \underbrace{1x}\ \underbrace{xx}\ \underbrace{xx}\ \underbrace{xx}$$

and the feedback connections in the carry-save tree are not used. The third multiplier, of length 28 bits, generates nine partial products, requiring the use of the feedback connection in the carry-save tree.

This can be avoided by additional truncation of the multiplier. To the 14 leading identical bits, we may add nine bits (for a total of 23 bits), and still generate only six partial products:

$$\underbrace{0.1}\ 11\ 11\ 11\ 11\ 11\ 11\ \underbrace{1x}\ \underbrace{xx}\ \underbrace{xx}\ \underbrace{xx}\ \underbrace{xx}$$

This, however, results in a new denominator guaranteed to have only $14 + 9 = 23$ leading identical bits, instead of 28 (the proof of this is left as an exercise for the reader). The next multiplier will therefore have only 23 leading identical bits, and again we may add nine extra bits without requiring the use of the feedback connections. This multiplier will result in a denominator with $23 + 9 = 32$ leading identical bits. Forming the two's complement of this denominator will give us the next multiplier with 32 leading identical bits, which can increase the number of leading identical bits in the denominator up to 64 and achieve convergence. Since this is the last multiply operation within the division, we can afford to use the feedback connections, so there is no need to limit the number of multiplier bits and all available 56 bits can be used. The sequence of multiplication factors now contains five multipliers of lengths of 10, 14, 23, 32, and 56 bits, increasing the number of multiply operations from 7 to 9. However, all these multiplications can be overlapped, leading to a total execution time of 18 clock cycles [1].   □

8.2 DIVISION BY RECIPROCATION

A somewhat different approach to performing division through multiplication is to first calculate the reciprocal of the divisor D and then multiply it by the dividend to form the final quotient. The reciprocal of D can be calculated using the Newton-Raphson iteration method. This is a method of finding the *zero* of a given function $f(x)$, where a zero of $f(x)$ is the solution of $f(x) = 0$. Let x_0 be the first approximation and let x_i be the estimate for the zero at the ith step. The next estimate, x_{i+1}, is calculated from

$$x_{i+1} = x_i - \frac{f(x_i)}{f'(x_i)} \tag{8.10}$$

where $f'(x)$ is the derivative of $f(x)$ with respect to x. For the function $f(x) = 1/x - D$, which has a zero at $x = 1/D$, $f'(x) = -1/x^2$, yielding

$$x_{i+1} = x_i(2 - D \cdot x_i). \tag{8.11}$$

x_{i+1} converges to the reciprocal of D and the convergence of this scheme is quadratic. To prove this, let δ_i denote the error in the ith step; i.e., $\delta_i = 1/D - x_i$. Simple algebraic manipulations show that $\delta_{i+1} = D\delta_i^2$. If D is a normalized fraction (i.e., $1/2 \leq D < 1$), then $\delta_i \leq 1$ and the error decreases quadratically.

If the first approximation is $x_0 = 1$, then $x_1 = (2 - D)$, and

$$x_2 = (2 - D) \cdot [2 - D(2 - D)] = (2 - D) \cdot [1 + (D - 1)^2]. \tag{8.12}$$

Repeatedly substituting expression (8.11) results in

$$\begin{aligned} x_i &= (2 - D)(1 + (D - 1)^2)(1 + (D - 1)^4) \cdots (1 + (D - 1)^{2^i}) \\ &= (1 - (D - 1))(1 + (D - 1)^2)(1 + (D - 1)^4) \cdots (1 + (D - 1)^{2^i}) \end{aligned} \tag{8.13}$$

If D is a normalized fraction then $(1 - D)$ is a fraction y satisfying $0 < y \leq \frac{1}{2}$. Therefore, the binomial series in Equation (8.13) is identical to the one used in Equation (8.4) and it converges to

$$\lim_{i \to \infty} x_i = \frac{1}{1 + (D - 1)} = \frac{1}{D}. \tag{8.14}$$

As in Section 8.1, we may reduce the required number of steps by reading the first approximation out of a table, rather than setting x_0 equal to 1. This table, stored in a ROM, accepts the j most significant digits of D (except for the first digit, which is always 1), and produces an approximation to $\frac{1}{D}$. The range $[0.5, 1)$ is divided into 2^j intervals (of size $\Delta = 0.5 \cdot 2^{-j}$ each) and it can be shown that the optimum value of x_0 for the kth interval ($k = 1, 2, \cdots, 2^j$) is the reciprocal of the number corresponding to the middle point of the interval (see Exercise 7). This middle point is $\frac{1}{2} + (k - \frac{1}{2}) \cdot \Delta$, and thus

$$x_0(k) = \frac{2^{j+1}}{2^j + k - \frac{1}{2}}. \tag{8.15}$$

This stepwise approximation for $j = 2$ is shown in Figure 8.1.

Instead of the stepwise approximation to $1/D$, a piecewise linear approximation can be employed. Its generation is more complicated but its accuracy is higher [3].

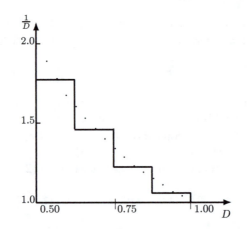

FIGURE 8.1   A stepwise approximation for the reciprocal $\frac{1}{D}$ (j = 2).

Example 8.4

The implementation of a division-by-reciprocal-approximation algorithm was proposed for the ZS-1 64-bit computer [5]. This computer uses the IEEE floating-point standard and, as a result, the significand d of the divisor is within the range $1 \leq d < 2$. The 15 most significant bits of the divisor (excluding the hidden bit), $1.d_1d_2\cdots d_{15}$, are used to address a ROM look-up table for the initial approximation x_0. The table is of size $32\mathrm{K} \times 16$ bits, producing an initial approximation in the range $0.5 \leq x_0 < 1$. x_0 therefore has the binary form $0.1y_2y_3\cdots y_{16}$. This approximation is calculated by taking the reciprocal of the mid-point between $1.d_1d_2\cdots d_{15}$ and its successor where the mid-point is $1.d_1d_2\cdots d_{15}1$. The reciprocal of the mid-point is rounded by adding 2^{-17}, and the result is truncated to yield the required 16 bits, $0.1y_2y_3\cdots y_{16}$.

The precision of this initial approximation was shown to be $|x_0 - 1/d| < 1.5 \cdot 2^{-16}$ [5]. Therefore, based on the quadratic convergence of the Newton-Raphson method, only two iterations are needed to achieve the required precision of 53 significand bits. Two iterations of the form of Equation (8.11) involve four multiplications and two complement operations. As was done in the implementation of the division-by-convergence scheme (in the IBM 360/91 system), additional simplifications were performed in order to reduce the overall execution time. In the first iteration that calculates $x_1 = x_0 \cdot (2 - d \cdot x_0)$, d and x_0 are multiplied and the complement $(2 - d \cdot x_0)$ is formed. To speed up the multiplication, the 16 bits of x_0 are multiplied by the 32 most significant bits of d, and the result is rounded to 32, instead of 48, bits. In the complement operation, only the one's complement is performed to avoid the carry-propagation in the two's complement calculation. This introduces an error of size 2^{-31}. x_1 is then computed by performing the multiplication $x_1 = x_0 \cdot (2 - dx_0)$ with 16 bits of x_0 multiplied by the 32 most significant bits of the multiplicand, forming a 32-bit product. The resulting x_1 is accurate to approximately 31 bits. In the second iteration, a similar set of operations is repeated to calculate the final approximation x_2. In $x_1 \cdot d$, only 64 bits are generated and then only the one's complement is calculated. Next, the multiplication $x_1(2 - d \cdot x_1)$ is performed producing the approximated value of $1/d$. This value is then multiplied by the dividend N and the resulting approximated quotient Q' is rounded according to any one of the four rounding schemes supported by the IEEE floating-point standard (round-to-nearest-even, round to 0, round to $\pm\infty$). This final rounding does not guarantee an accurately rounded result for all values of the operand d. $\square$

A major disadvantage of most implementations of the division-by-reciprocation algorithm is that the accuracy is smaller than that achieved by the add/subtract type of division algorithms described in Chapter 7. Corrective actions can be taken to guarantee that the least significant bit is correctly rounded. However, the additional computation usually slows down the division. Thus, when deciding between a division algorithm that is based on multiplications and an algorithm based on add/subtract operations, precision, speed, and cost trade-offs must be taken into account. The final decision depends on the technology available. The precision required by the IEEE standard has been achieved at a reasonable cost and speed in the IBM RISC/6000 [7]. A double-width data-path is used there to implement the division by reciprocation algorithm. All operations are done in a fused multiply-add unit (see Section 6.5) resulting in a double-length estimate Q' of the quotient. A remainder is then calculated (using the fused multiply-add),

$$R = N - D \times Q'.$$

This remainder is used to compute a properly rounded result (again using the fused multiply-add unit) in the desired rounding mode,

$$Q = Q' + R \times \frac{1}{D}$$

where $\frac{1}{D}$ is the result of the Newton-Raphson iterations.

A different solution is to estimate the error in Q' by calculating $N' = DQ'$ [6]. If Q' is sufficiently accurate (at least to $n+1$ bits where n is the number of bits in the significand), the least significant bits of N' provide the direction of the error in Q'. Based on this information and the desired rounding mode, Q' can be corrected by either adding or subtracting 1 at the $n+1$ bit position or by truncating it.

8.3 EXERCISES

8.1. Show that if a truncated multiplier is used then the denominator D_i will converge toward 1 from below or above.

8.2. Prove that a normalized denominator D_i in the division-by-convergence algorithm with a identical leading bits, when multiplied by its two's complement R_i with a identical leading bits and only b extra bits ($b \leq a$), will generate a new denominator D_{i+1} with at least $a + b$ leading identical bits.

8.3. Prove that a ROM of size $2^7 \times 10$ can always provide a multiplying factor R_0 such that $D_1 = D \cdot R_0$ has at least seven leading 1's when D is a normalized fraction ($0.5 \leq D < 1$) in the division-by-convergence algorithm. Can the size of this table be further reduced?

8.4. Evaluate the overall time needed to accumulate 28 partial products using a pipelined partial CSA tree as shown in Figure 5.31, capable of accepting six new partial products at once. Compare it to the time needed if no overlapping is allowed.

8.5. Prove that $\delta_{i+1} = D\delta_i^2$ in the division-by-reciprocation algorithm.

8.6. A different representation of the algorithm for calculating the reciprocal is obtained by introducing a new variable, z_i, which equals $z_i = Dx_i$. Show that the resulting equation for z_i is

$$z_{i+1} = z_i(2 - z_i)$$

Show also that from z_i we can calculate x_i using the equation $x_{i+1} = x_i(2 - z_i)$. If the initial value for x_i is $x_0 = 1$, then the initial value for z_i is $z_0 = D$. When x_i converges to $1/D$, z_i converges to 1. Thus, the comparison of z_i to 1 allows us to determine whether convergence has been achieved. Write the above scheme as a ratio between x_0 and z_0 similar to Equation (8.1). What will be the form of the multiplication factors R_i?

8.7. Show that the optimal initial value x_0 for the kth interval ($k = 1, 2, \cdots, 2^j$) in the division-by-reciprocation algorithm is the reciprocal of the number corresponding to its middle point.

8.8. To calculate the square root of a given operand N it has been suggested to apply the Newton-Raphson method with $f(x) = 1 - 1/(Nx^2)$. Show the corresponding iteration rule and prove that the convergence is quadratic. Apply this procedure to calculate $\sqrt{3}$ to five decimal places.

8.9. Given an operand A that is a normalized fraction ($0.5 \leq A < 1$), what function $z = f(A)$ will the following procedure calculate?
(*i*) $x_{i+1} = x_i \cdot r_i^2$ with $x_0 = A$.
(*ii*) $z_{i+1} = z_i \cdot r_i$ with $z_0 = A$.
with $r_i = 1 + (1 - x_i)/2$.
How many iterations are needed in this calculation and what operations are executed in each iteration?
How is the multiplying coefficient calculated?

Estimate the error in the final result.

Can you suggest ways to speed up the calculation?

Given $A = 0.11100001$, calculate the 8-bit result $z = f(A)$ using the above procedure and compare it to the correct result.

8.4 REFERENCES

[1] F. S. ANDERSON *et al.*, "The IBM system/360 model 91: Floating-point execution unit," *IBM Journal Res. and Dev., 11* (Jan. 1967), 34-53.

[2] M. D. ERCEGOVAC, T. LANG, J.-M. MULLER, and A. TISSERAND, "Reciprocation, square root, inverse square root, and some elementary functions using samll multipliers," *IEEE Trans. on Computers, 49* (July 2000), 628-637.

[3] D. FERRARI, "A division method using a parallel multiplier," *IEEE Trans. on Computers, EC-16* (April 1967), 224-226.

[4] M. J. FLYNN, "On division by functional iteration," *IEEE Trans. on Computers, C-19* (Aug. 1970), 702-706.

[5] D. L. FOWLER and J. E. SMITH, "An accurate high speed implementation of division by reciprocal approximation," *Proc. of 9th Symp. on Computer Arithmetic* (1989), 60-67.

[6] H. KABUO *et al.*, "Accurate rounding scheme for the Newton-Raphson method using redundant binary arithmetic," *IEEE Trans. on Computers, 43* (January 1994), 43-51.

[7] P. W. MARKSTEIN, "Computation of elementary functions on the IBM RISC system/6000 processor," *IBM Journal Res. and Dev., 34* (Jan. 1990), 111-119.

[8] S. F. OBERMAN, and M. J. FLYNN, "Division algorithms and implementations," *IEEE Trans. on Computers, 46* (August 1997), 833-854.

[9] P. SODERQUIST, and M. LEESER, "Area and performance tradeoffs in floating-point divide and square-root implementations," *ACM Computing Surveys, 28* (September 1996), 518-564.

[10] C. S. WALLACE, "A suggestion for a fast multiplier," *IEEE Trans. Elect. Comp., EC-13* (Feb. 1964), 14-17.

9

EVALUATION OF
ELEMENTARY FUNCTIONS

Our objective in this chapter is to find efficient algorithms for evaluating elementary functions (like e^x, $\ln x$, $\sin x$, $\cos x$, etc.) in hardware. These algorithms should be accurate and sufficiently fast in comparison to alternative algorithms implemented in software. Such algorithms are especially useful when designing scientific hand-held calculators and numerical (usually floating-point) processors ([12]).

One straightforward method is the use of a look-up table. To evaluate $y = F(x)$ where x and y are numbers of length n bits, a ROM of size $2^n \times n$ is required. For $n \geq 20$, this size is prohibitive ($\geq 16M$). Another method is to use a Taylor series expansion; e.g.,

$$e^x = \sum_{i=0}^{\infty} \frac{x^i}{i!}. \tag{9.1}$$

This series converges rapidly for a fractional argument x. However, for x close to 1, a large number of steps is needed. In addition, its hardware implementation is complex, since separate logical networks are needed for different elementary functions. Even for software implementations there exist more efficient algorithms requiring a smaller number of computational steps for the same precision. The most commonly used are based on either polynomial or rational approximations.

Consider, for example, the exponential function e^x. We can express it in the form $e^x = 2^{x \log_2 e}$ and partition the exponent $x \log_2 e$ into its integer part I and its fractional part f, i.e., $x \log_2 e = I + f$. Thus,

$$e^x = 2^I \cdot 2^f. \tag{9.2}$$

225

The incorporation of the factor 2^I in either fixed-point or floating-point calculations is straightforward. To evaluate 2^f one can use a rational approximation, which is the ratio between a numerator polynomial and a denominator polynomial. One such approximation is based on two degree-5 polynomials

$$2^f = \frac{((((a_5 f + a_4)f + a_3)f + a_2)f + a_1)f + a_0}{((((b_5 f + b_4)f + b_3)f + b_2)f + b_1)f + 1} \tag{9.3}$$

where a_i and b_i $(i = 0, 1, \cdots, 5)$ are known constants [7]. To evaluate this approximation we need 10 multiplications, 10 additions, and 1 division. Other approximations requiring a smaller execution time are available.

9.1 THE EXPONENTIAL FUNCTION

There are alternative methods for evaluating these elementary functions that are more readily implemented in hardware. Most existing algorithms for evaluating elementary functions are similar to the division-by-convergence algorithm (see Chapter 8). They involve two (or more) recursive formulas related in such a way that when one formula is forced to a constant value the other formula yields the desired result.

For example, to evaluate the exponential function $y = e^{x_0}$ for a fractional argument x_0, we may use the formulas

$$
\begin{array}{rll}
(i) & x_{i+1} & = & x_i - \ln b_i, \\
(ii) & y_{i+1} & = & y_i \cdot b_i.
\end{array}
\tag{9.4}
$$

The b_i's are selected in such a way that the elements of the sequence x_0, x_1, $\cdots$, x_m approach 0 where m is the number of iterations needed to ensure the convergence to zero, i.e., $x_m = 0$. As will become evident later, the convergence is linear. In other words, m is a linear function of the number of bits, n.

To find the value approached by the corresponding sequence y_0, y_1, $\cdots$, y_m note that [6]

$$y_{i+1} \cdot e^{x_{i+1}} = y_i \cdot b_i \cdot e^{x_i - \ln b_i} = y_i \cdot e^{x_i}. \tag{9.5}$$

In particular,

$$y_m \cdot e^{x_m} = y_0 \cdot e^{x_0}, \tag{9.6}$$

and since $x_m = 0$, y_m yields the desired result,

$$y_m = y_0 \cdot e^{x_0}. \tag{9.7}$$

The similarity between the above algorithm for calculating e^{x_0} and the division-by-convergence algorithm is now apparent. In the division-by-convergence algorithm the ratio N_i/D_i is kept constant while here the product $y_i \cdot e^{x_i}$ remains constant independent of the specific values of the b_i's.

For the above method to be efficient we need a simple way of selecting the b_i's to ensure the convergence of x_{i+1} to 0. Also, the multiplication by b_i should be simple and not overly time-consuming.

To simplify the multiplication, the b_i's are given the form $b_i = (1 + s_i 2^{-i})$, where $s_i \in \{-1, 0, 1\}$. This way, the multiplication is reduced to shift and add operations. Also, the term $\ln b_i = \ln(1 + s_i 2^{-i})$ in formula (i) is either positive or negative. This is necessary in order to ensure the convergence of either a positive or negative x_{i+1} to 0. Clearly, all possible values of $\ln(1 \pm 2^{-i})$ must be precalculated, since their online calculation will slow down the execution enough to render the algorithm useless. These precalculated quantities are stored in a look-up table; e.g., a ROM of size $2n \times n$.

Substituting $b_i = (1 + s_i 2^{-i})$ in Equation (9.4), we obtain

$$
\begin{aligned}
(i) \quad & x_{i+1} & = & \quad x_i - \ln(1 + s_i 2^{-i}), \\
(ii) \quad & y_{i+1} & = & \quad y_i \cdot (1 + s_i 2^{-i}).
\end{aligned} \tag{9.8}
$$

To calculate the exponential function we have to find the vector $\mathbf{s} = \{s_0, s_1, \cdots, s_{m-1}\}$ so that

$$
x_m = 0 \quad \text{or} \quad \sum_{l=0}^{m-1} \ln(1 + s_l 2^{-l}) = x_0.
$$

If s_l is restricted to $\{-1, 0, 1\}$, then the smallest and largest values of x_0 that can be represented in such a way correspond to all $s_l = -1$ and all $s_l = 1$, respectively, and consequently,

$$
\sum_{l=1}^{m-1} \ln(1 - 2^{-l}) \ \leq \ x_0 \ \leq \ \sum_{l=0}^{m-1} \ln(1 + 2^{-l}). \tag{9.9}
$$

For every x_0 in this interval there exists a vector $\mathbf{s}$ such that the convergence of x_m to 0 is guaranteed. For large enough values of m $(m \geq 20)$ the bounds in inequality (9.9) converge, yielding the following domain for x_0:

$$
-1.24 \leq x_0 \leq 1.56 \tag{9.10}
$$

If we restrict the argument x_0 to positive fractions, we may use a simple scheme for selecting s_i. This is a one-sided selection rule; i.e., $s_i \in \{0, 1\}$, similar to the quotient-bit selection rule in the restoring division algorithm. In step $(i+1)$ we form the difference $D = x_i - \ln(1 + 2^{-i})$. If D is positive or zero, we set $s_i = 1$ and $x_{i+1} = D$; if D is negative, we set $s_i = 0$ and $x_{i+1} = x_i$.

We may analyze the possible outcomes of the subtract operation in formula (i) by examining the Taylor series expansion of $\ln(1 + s_i 2^{-i})$:

$$
\ln(1 + s_i 2^{-i}) \ = \ s_i 2^{-i} \ - \ \frac{(s_i)^2 \ 2^{-2i}}{2} \ + \ s_i \frac{2^{-3i}}{3} \ - \ \cdots \tag{9.11}
$$

i	$1 + 2^{-i}$	$\ln(1 + 2^{-i})$	$1 - 2^{-i}$	$\ln(1 - 2^{-i})$
0	10.00000 00000	0.10110 00110	0	—
1	1.10000 00000	0.01100 11111	0.10000 00000	$-.10110\,00110$
2	1.01000 00000	0.00111 00100	0.11000 00000	$-.01001\,00111$
3	1.00100 00000	0.00011 11001	0.11100 00000	$-.00100\,01001$
4	1.00010 00000	0.00001 11110	0.11110 00000	$-.00010\,00010$
5	1.00001 00000	0.00001 00000	0.11111 00000	$-.00001\,00001$
6	1.00000 10000	0.00000 10000	0.11111 10000	$-.00000\,10000$
7	1.00000 01000	0.00000 01000	0.11111 11000	$-.00000\,01000$
8	1.00000 00100	0.00000 00100	0.11111 11100	$-.00000\,00100$
9	1.00000 00010	0.00000 00010	0.11111 11110	$-.00000\,00010$
10	1.00000 00001	0.00000 00001	0.11111 11111	$-.00000\,00001$

TABLE 9.1   The value of $\ln(1 \pm 2^{-i})$ with 10-bit precision.

At step i we can cancel the bit whose weight is 2^{-i} by subtracting $\ln(1 + s_i 2^{-i})$ from x_i. This implies that we may need up to n steps to ensure the convergence of an n-bit fraction to zero. The convergence is therefore linear and $m = n$. We can rely on this fact and slightly modify the selection rule to improve the performance of the algorithm. If the bit in the ith position is 1, we select $s_i = 1$ and calculate $x_{i+1} = x_i - \ln(1 + s_i 2^{-i})$. If the bit is 0, we select $s_i = 0$ and go on to the next step without performing any subtraction. As a result, we have the capability of skipping over zeros. More complex selection rules can be used as well [3].

Example 9.1

To calculate $e^{0.25}$ in 10-bit precision we need a table of $\ln(1 \pm 2^{-i})$ for $i = 0, 1, 2, \cdots, 10$ with each entry having 10 fractional bits. The required entries have been calculated and are shown in Table 9.1. The calculated entries have been rounded-to-nearest rather than truncated. As a result, we obtain $\ln(1 \pm 2^{-i}) = \pm 2^{-i}$ for $i \geq 6$.

Suppose that we use the one-sided selection rule. Starting with $x_0 = 0.25_{10} = 0.01000\,000000_2$, we must select $s_0 = s_1 = 0$, since $x_0 < \ln(1 + 2^{-1})$, and thus $x_2 = x_0$. We then select $s_2 = 1$ and obtain $x_3 = x_2 - 0.00111\,00100 = 0.00000\,11100$. We then set $s_3 = s_4 = 0$, and at this point it becomes evident that we must select $s_5 = 0$, $s_6 = s_7 = s_8 = 1$, and $s_9 = s_{10} = 0$, yielding $x_{11} = 0$. In parallel, we calculate $y_{11} = y_0 \cdot (1 + 2^{-2})(1 + 2^{-6})(1 + 2^{-7})(1 + 2^{-8})$ yielding $y_{11} = 1.01001\,00011_2 = 1.28418_{10}$. The exact value with a precision of five decimal digits is 1.28403_{10}. The approximation error equals $0.158 \cdot 2^{-10}$. The exact steps of the calculation are summarized in the table at the top of the next page:

If we allow s_i to assume the value -1, then the selected values are $(s_0, s_1, s_2, s_3, s_4, s_5, s_6, s_7, s_8, s_9, s_{10}) = 0, 1, -1, 1, 0, 0, 0, 1, 1, 1, 1$ resulting in $x_{11} = 0$ and $y_{11} = 1.01001\,00010_2$. The approximation error equals $0.842 \cdot 2^{-10}$.

i	x_i	y_i	s_i
0	0.01000 00000	1.00000 00000	0
1	0.01000 00000	1.00000 00000	0
2	0.01000 00000	1.00000 00000	1
3	0.00000 11100	1.01000 00000	0
4	0.00000 11100	1.01000 00000	0
5	0.00000 11100	1.01000 00000	0
6	0.00000 11100	1.01000 00000	1
7	0.00000 01100	1.01000 10100	1
8	0.00000 00100	1.01000 11110	1
9	0.00000 00000	1.01001 00011	0
10	0.00000 00000	1.01001 00011	0
11	0.00000 00000	1.01001 00011	

$\square$

The domain for the argument of the exponential function can be extended by writing the argument as $x = x \log_2 e \cdot \ln 2$, then partitioning the first term into its integer part and its fractional part; i.e., $x \log_2 e = I + f$, where I is the integer part and f is a fraction ($0 \leq f < 1$). We need, therefore, to calculate

$$y = e^x = e^{(I+f)\ln 2} = e^{I \ln 2 + f \ln 2} = 2^I \cdot e^{f \ln 2}.$$

We set $x_0 = f \ln 2$, and consequently $0 \leq x_0 < \ln 2 = 0.693$. The factor 2^I is easily dealt with, either by incorporating it into the exponent part of the floating-point number, or through a shift operation if fixed-point arithmetic is used.

9.2 THE LOGARITHM FUNCTION

The procedure for calculating e^x in the previous section is based on continued summation of terms of the form $\ln(1 + s_l 2^{-l})$ to force the convergence of x_i to 0. This type of procedure is called *additive normalization*. In a similar way, we may define a *multiplicative normalization* procedure as one in which x_i is forced to 1 (or some other nonzero constant) by continued multiplication with precalculated factors.

The set of recursive formulas for multiplicative normalization has the following general form:

$$
\begin{aligned}
(i) \quad & x_{i+1} = x_i \cdot b_i, \\
(ii) \quad & y_{i+1} = y_i - g(b_i).
\end{aligned}
\tag{9.12}
$$

b_i is selected so that x_{i+1} approaches 1, i.e., $x_{i+1} = x_0 \prod_{l=0}^{i} b_l \to 1$ and thus, after m steps

$$\prod_{l=0}^{m-1} b_l = \frac{1}{x_0} \quad \text{(the multiplicative inverse).} \tag{9.13}$$

The multiplication factor b_i is again given the form $(1 + s_i 2^{-i})$, and for $s_i \in \{-1, 0, 1\}$ the following inequality determines the domain of the algorithm:

$$\prod_{l=1}^{m-1} (1 - 2^{-l}) \leq \prod_{l=0}^{m-1} (1 + s_l 2^{-l}) \leq \prod_{l=0}^{m-1} (1 + 2^{-l}) \tag{9.14}$$

These bounds can be calculated from the corresponding bounds in inequality (9.9) and, for large enough values of m, they converge to

$$0.29 \leq \prod_{l=0}^{m-1} (1 + s_l 2^{-l}) \leq 4.77.$$

Consequently, the domain for x_0 is

$$0.21 \leq x_0 \leq 3.45. \tag{9.15}$$

Therefore, positive normalized fractions are within the domain.

If the argument for the logarithm function is not already represented as a normalized floating-point number, we may rewrite it as such, obtaining $x = x_0 \cdot 2^{E_x}$, where $0.5 \leq x_0 < 1$. Thus, $\ln x = \ln x_0 + E_x \ln 2$.

A simple one-sided selection rule with $s_i \in \{0, 1\}$ can be employed here as well. If x_i already has i leading 1's (i.e., $x_i = 0.\underbrace{11 \cdots 1}_{i} 0zz \cdots z$), then the

multiplication

$$x_{i+1} = x_i b_i = x_i + x_i s_i 2^{-i}$$

with $s_i = 1$ will produce $(i + 1)$ leading 1's in x_{i+1}.

Formula (ii) in Equation (9.12) results in $y_{i+1} = y_0 - \sum_{l=0}^{i} g(b_l)$. If we select $g(b_l) = \ln b_l$, then

$$y_{i+1} = y_0 - \sum_{l=0}^{i} \ln b_l = y_0 - \ln \prod_{l=0}^{i} b_l$$

Therefore, when x_{i+1} approaches 1, y_{i+1} converges to

$$y_m = y_0 - \ln \frac{1}{x_0} = y_0 + \ln x_0. \tag{9.16}$$

Thus, the multiplicative normalization algorithm in (9.12) with $g(b_i) = \ln(1 + s_i 2^{-i})$ can be employed to calculate the natural logarithm function $F(x_0) = \ln x_0$. Notice that the same table of $\ln(1 + s_i 2^{-i})$ that was used for evaluating the exponential function is needed here. Also note that we may replace the logarithm base e by any other base. Of particular interest are the base 10 logarithm and exponential functions. These are useful in hand-held calculators.

Finally note that, in a manner analogous to Equation (9.5), we may write

$$y_{i+1} + \ln x_{i+1} = (y_i - \ln b_i) + (\ln x_i + \ln b_i) = y_i + \ln x_i, \qquad (9.17)$$

indicating that y_{i+1} approaches $y_0 + \ln x_0$ when x_{i+1} approaches 1.

Example 9.2

To calculate $\ln(0.50)$ in 10-bit precision we use the entries of Table 9.1. The steps of the calculation are shown in the table below:

i	x_i	y_i	s_i
0	0.10000 00000	0.00000 00000	0
1	0.10000 00000	0.00000 00000	1
2	0.11000 00000	−0.01100 11111	1
3	0.11110 00000	−0.10100 00011	0
4	0.11110 00000	−0.10100 00011	1
5	0.11111 11100	−0.10110 00001	0
6	0.11111 11100	−0.10110 00001	0
7	0.11111 11100	−0.10110 00001	0
8	0.11111 11100	−0.10110 00001	1
9	1.00000 00000	−0.10110 00101	0
10	1.00000 00000	−0.10110 00101	0
11	1.00000 00000	−0.10110 00101	

Thus, $y_{11} = -0.10110\,00101_2 = -0.69238_{10}$. The exact result (in five decimal digits) is -0.69315 and the approximation error is $0.783 \cdot 2^{-10}$. □

For multiplicative normalization we can also consider the case where the operation in formula (ii), Equation (9.12), is multiplication instead of subtraction. Selecting, for example, $g(b_i) = b_i^k$ (where k is any real number) formula (ii) yields

$$y_m = y_0 \cdot \prod_{l=0}^{m-1} g(b_l) = y_0 \cdot \left(\prod_{l=0}^{m-1} b_l \right)^k \;\rightarrow\; y_0 \cdot \left(\frac{1}{x_0} \right)^k. \qquad (9.18)$$

For example, if $k = 1$, then $y_{i+1} \rightarrow y_0/x_0$ and a divide operation is executed. If $k = 1/2$, $y_{i+1} \rightarrow y_0/\sqrt{x_0}$ and the reciprocal of the square root is calculated, and if we set $y_0 = x_0$, we obtain $y_{i+1} \rightarrow \sqrt{x_0}$. Here $g(b_i) = \sqrt{b_i}$, and to simplify the multiplication in formula (i) we may select $g(b_i) = (1 + s_i 2^{-i})$. Consequently,

$$b_i = (1 + s_i 2^{-i})^2 = 1 + 2 s_i 2^{-i} + s_i^2 2^{-2i} = 1 + s_i 2^{-i+1} + s_i^2 2^{-2i}. \qquad (9.19)$$

The multiplication by b_i in formula (ii) is now more complex and as a result, the efficiency of the method described above for square root extraction is questionable.

9.3 THE TRIGONOMETRIC FUNCTIONS

To evaluate the trigonometric functions we use the well-known relation between the exponential function and the trigonometric functions

$$e^{jx} = \cos x + j \sin x \qquad (9.20)$$

where $j = \sqrt{-1}$. For the exponential function we used the recursive formulas for additive normalization, which have the following general form:

$$\begin{align}
(i) \quad x_{i+1} &= x_i - g(b_i), \\
(ii) \quad y_{i+1} &= y_i \cdot b_i.
\end{align} \qquad (9.21)$$

To calculate the trigonometric functions we select $b_i = (1 + js_i 2^{-i})$. This complex number can be written in the form

$$\sqrt{1^2 + (s_i 2^{-i})^2} \cdot e^{j\theta_i} = \sqrt{1 + (s_i 2^{-i})^2} \cdot (\cos \theta_i + j \sin \theta_i)$$

where $\theta_i = \tan^{-1}(s_i 2^{-i})$. This is the polar form of a complex number which, in general, is

$$A + jB = \sqrt{A^2 + B^2} \cdot e^{j\theta} = \sqrt{A^2 + B^2} \cdot (\cos \theta + j \sin \theta) \qquad (9.22)$$

where $\theta = \tan^{-1} \frac{B}{A}$.

The s_i's are selected so that the elements of the sequence $x_0, x_1, \cdots, x_m$ approach 0. To find the value approached by the corresponding sequence $y_0, y_1, \cdots, y_m$ we form an expression analogous to that in Equation (9.5):

$$y_{i+1} \cdot e^{jx_{i+1}} = y_i \cdot b_i \cdot e^{jx_i - jg(b_i)} = y_i \cdot e^{jx_i} \sqrt{1 + (s_i 2^{-i})^2} \cdot e^{j\theta_i} e^{-jg(b_i)}$$

Setting

$$g(b_i) = \theta_i = \tan^{-1}(s_i 2^{-i}) \qquad (9.23)$$

yields

$$y_{i+1} \cdot e^{jx_{i+1}} = y_i \cdot e^{jx_i} \sqrt{1 + (s_i 2^{-i})^2}. \qquad (9.24)$$

Note that if it were not for the term $\sqrt{1 + (s_i 2^{-i})^2}$ the product $y_i \cdot e^{jx_i}$ would have been a constant. Equation (9.24) results in

$$y_m \cdot e^{jx_m} = y_0 \cdot e^{jx_0} \prod_{l=0}^{m-1} \sqrt{1 + (s_l 2^{-l})^2}.$$

Denote $K = \prod_{l=0}^{m-1} \sqrt{1 + (s_l 2^{-l})^2}$; then for $x_m = 0$ we obtain

$$y_m = y_0 \cdot K \cdot e^{jx_0} = y_0 \cdot K \cdot (\cos x_0 + j \sin x_0). \qquad (9.25)$$

The convergence domain for this algorithm for large enough values of m $(m \geq 20)$ is

$$-1.743 = \sum_{l=0}^{m-1} \tan^{-1}(-2^{-l}) \leq x_0 \leq \sum_{l=0}^{m-1} \tan^{-1}(2^{-l}) = 1.743 \qquad (9.26)$$

which includes the useful domain $0 \leq x_0 \leq \pi/2 = 1.57$.

The x_i's are real numbers, but the y_i's are complex numbers. To separate the real and imaginary parts of y_i, let y_i equal $Z_i + jW_i$ where Z_i is the real part of y_i and W_i is its imaginary part. Formula (ii) now takes the form

$$(ii) \; y_{i+1} = Z_{i+1} + jW_{i+1} = (Z_i + jW_i)(1 + js_i 2^{-i}). \qquad (9.27)$$

Therefore, the recursive formulas for Z_{i+1} and W_{i+1} are

$$\begin{array}{rcl} (ii) & Z_{i+1} & = & Z_i - s_i 2^{-i} \cdot W_i, \\ (ii)' & W_{i+1} & = & W_i + s_i 2^{-i} \cdot Z_i. \end{array} \qquad (9.28)$$

The initial value y_0 can also, in general, be a complex number, $y_0 = Z_0 + jW_0$, but if we set W_0 equal to 0, making Z_0 equal to y_0, we obtain the desired values, which according to Equation (9.25) are

$$\begin{array}{rcl} Z_m & = & y_0 \cdot K \cdot \cos x_0, \\ W_m & = & y_0 \cdot K \cdot \sin x_0. \end{array} \qquad (9.29)$$

The recursive formulas for the calculation of $\sin x$ and $\cos x$ were developed somewhat differently by Volder [17] through rotations in a polar system. The resulting implementation was named CORDIC (COordinated Rotation DIgital Computer). This formulation was later extended to include division and the evaluation of hyperbolic functions [18].

A possible selection rule for s_i to ensure the convergence of x_{i+1} to 0 when x_0 is in the domain presented in Equation (9.26) is

$$s_i = \begin{cases} 1 & \text{if} \;\; x_i > \tan^{-1}(2^{-i}) \\ 0 & \text{if} \;\; |x_i| \leq \tan^{-1}(2^{-i}) \\ -1 & \text{if} \;\; x_i < -\tan^{-1}(2^{-i}). \end{cases} \qquad (9.30)$$

Since $\tan^{-1}(-\alpha) = -\tan^{-1}(\alpha)$, $\tan^{-1}(s_i 2^{-i}) = s_i \tan^{-1}(2^{-i})$, and thus only n constants, rather than $2n$, must be stored in a ROM. With this set of values for s_i, $\{-1, 0, 1\}$, we obtain

$$K = \prod_{i=0}^{m-1} \sqrt{1 + (s_i 2^{-i})^2}. \tag{9.31}$$

Therefore, K is not a constant but depends on the vector $\mathbf{s}$, which in turn depends upon x_0. In addition, the above selection rule requires a full-length comparison between x_i and $\tan^{-1}(2^{-i})$. If we restrict s_i to assume only the values $\{-1, 1\}$, then

$$K = \prod_{i=0}^{m-1} \sqrt{1 + 2^{-2i}},$$

and it is a constant that can be precalculated. For $m > 16$, $K = 1.6468$. We may then set $y_0 = 1/K$ and, as a result, obtain $Z_m = \cos x_0$ and $W_m = \sin x_0$.

The selection rule for s_i now becomes

$$s_i = \begin{cases} 1 & \text{if } x_i \geq 0 \\ -1 & \text{if } x_i < 0. \end{cases} \tag{9.32}$$

This is similar to the selection rule for nonrestoring division, and a full-length comparison is not needed. To examine the rate of convergence of x_{i+1} to 0, consider the Taylor series expansion of $\tan^{-1}(s_i 2^{-i})$ (in radians):

$$\tan^{-1}(s_i 2^{-i}) = s_i 2^{-i} - s_i \frac{2^{-3i}}{3} + \cdots \tag{9.33}$$

Again, linear convergence is achieved. Also, for $i > n/3$, all terms except the first in the above series are negligible and, as a result, a considerable reduction in the size of the ROM is possible.

i	2^{-i}	$\arctan(2^{-i})$
0	1.0000000000	0.1100100100
1	0.1000000000	0.0111011011
2	0.0100000000	0.0011111011
3	0.0010000000	0.0001111111
4	0.0001000000	0.0001000000
5	0.0000100000	0.0000100000
6	0.0000010000	0.0000010000
7	0.0000001000	0.0000001000
8	0.0000000100	0.0000000100
9	0.0000000010	0.0000000010
10	0.0000000001	0.0000000001

TABLE 9.2   The value of $\arctan(2^{-i})$ with 10-bit precision.

Example 9.3

To calculate $\sin(\pi/4)$ in 10-bit precision we need a table of $\arctan(2^{-i})$ for $i = 0, 1, 2, \cdots, 10$ with each entry having 10 fractional bits. The required entries have been calculated and are shown in Table 9.2. The calculated entries have been rounded-to-nearest rather than truncated. As a result, we obtain $\arctan(2^{-i}) = 2^{-i}$ for $i \geq 4$.

The input operand is $\pi/4 = 0.785398_{10} = 0.1100100100_2$. Initially we set $Z_0 = 1/K$ which in 10-bit precision is $0.1001101110_2 (= 0.6074_{10})$. We then set $s_i = sign(x_i)$ per Equation (9.32). The final results are $Z_{11} = 0.1011010011_2$ and $W_{11} = 0.1011010100_2$ while the "exact" results in 10-bit precision are $0.1011010100_2 = 0.7071_{10}$ and $0.1011010100_2 = 0.7071_{10}$, respectively. The approximation error is thus not larger than 2^{-10}. The exact steps of the calculation are summarized in the table below.

i	x_i	Z_i	W_i	s_i
0	0.1100100100	0.1001101110	0.0000000000	1
1	0.0000000000	0.1001101110	0.1001101110	1
2	−0.0111011011	0.0100110111	0.1110100101	−1
3	−0.0011100000	0.1000100000	0.1101010111	−1
4	−0.0001100001	0.1010001011	0.1100010011	−1
5	−0.0000100001	0.1010111100	0.1011101010	−1
6	−0.0000000001	0.1011010011	0.1011010100	−1
7	0.0000001111	0.1011011110	0.1011001001	1
8	0.0000000111	0.1011011000	0.1011001111	1
9	0.0000000011	0.1011010101	0.1011010010	1
10	0.0000000001	0.1011010100	0.1011010011	1
11	0.0000000000	0.1011010011	0.1011010100	

$\square$

9.4 THE INVERSE TRIGONOMETRIC FUNCTIONS

To evaluate the inverse trigonometric functions we use the multiplicative normalization algorithm for the inverse exponential function; i.e., the function $\ln x$ described in Section 9.2. The general form of multiplicative normalization is

$$
\begin{aligned}
(i) \quad x_{i+1} &= x_i \cdot b_i, \\
(ii) \quad y_{i+1} &= y_i + g(b_i).
\end{aligned}
\tag{9.34}
$$

If we set $b_i = (1 + j s_i 2^{-i})$ as in Section 9.2, then formula (i) yields

$$
x_m = x_0 \cdot \prod_{l=0}^{m-1} b_l = x_0 \cdot \prod_{l=0}^{m-1} (1 + j s_l 2^{-l}) = x_0 \cdot K e^{j\theta},
\tag{9.35}
$$

where

$$\theta = \sum_{l=0}^{m-1} \theta_l, \theta_l = \tan^{-1}(s_l 2^{-l}) \quad \text{and} \quad K = \prod_{l=0}^{m-1} \sqrt{1 + (s_l 2^{-l})^2}.$$

Similarly, if we set $g(b_i) = \theta_i = \tan^{-1}(s_i 2^{-i})$ and $y_0 = 0$ then formula (ii) results in

$$y_m = \sum_{l=0}^{m-1} \theta_l = \theta. \tag{9.36}$$

Here, x_i is a complex variable in the form $U_i + jV_i$. To obtain the recursive formulas for the real and imaginary parts, consider again formula (i):

$$(i)\ U_{i+1} + jV_{i+1} = (U_i + jV_i) \cdot (1 + js_i 2^{-i})$$

Separating the real and imaginary parts, we obtain

$$\begin{aligned}
(i)\quad U_{i+1} &= U_i - s_i 2^{-i} \cdot V_i, \\
(i)'\quad V_{i+1} &= V_i + s_i 2^{-i} \cdot U_i.
\end{aligned} \tag{9.37}$$

To find the relationship between the values of U_m, V_m and y_m we form the following expression:

$$x_{i+1} \cdot e^{-jy_{i+1}} = x_i \cdot b_i \cdot e^{-jy_i} e^{-jg(b_i)}$$

Substituting the selected values of b_i and $g(b_i)$ yields

$$x_{i+1} \cdot e^{-jy_{i+1}} = x_i \cdot e^{-jy_i} \cdot \sqrt{1 + (s_i 2^{-i})^2}, \tag{9.38}$$

resulting in

$$x_m \cdot e^{-jy_m} = x_m \cdot e^{-j\theta} = x_0 \cdot K, \tag{9.39}$$

where

$$K = \prod_{l=0}^{m-1} \sqrt{1 + (s_l 2^{-l})^2}$$

as in Section 9.3.

Replacing x_m and x_0 in Equation (9.39) by the corresponding real and imaginary parts yields

$$(U_m + jV_m) \cdot (\cos\theta - j\sin\theta) = K \cdot (U_0 + jV_0). \tag{9.40}$$

Thus,

$$\begin{aligned}
U_m \cdot \cos\theta + V_m \cdot \sin\theta &= K \cdot U_0, \\
-U_m \cdot \sin\theta + V_m \cdot \cos\theta &= K \cdot V_0.
\end{aligned} \tag{9.41}$$

Setting $V_0 = 0$ and $U_0 = 1/K$, we obtain

$$U_m = \cos\theta \quad \text{and} \quad V_m = \sin\theta.$$

Recall that K is a constant if we restrict s_i to the values $\{-1, 1\}$. If, for a given argument C, we need to calculate $\phi = \cos^{-1} C$, we select the s_i's in a way that $U_m \to C$ and, as a result, $\theta \to \phi$. The angle θ is obtained from $y_m = \theta$, where

$$\theta = \sum_{l=0}^{m-1} \tan^{-1}(s_l 2^{-l}) = \sum_{l=0}^{m-1} s_l \tan^{-1}(2^{-l}).$$

Consequently, a table of $\tan^{-1}(2^{-i})$ is needed, as it is for sine and cosine. Similarly, to calculate $\phi = \sin^{-1} D$ for a given D we select the s_i's so that $V_m \to D$ and again $\theta \to \phi$.

To evaluate the inverse tangent, $\tan^{-1} C$, let $V_m \to 0$, and from Equation (9.41) we obtain $\tan\theta = -\frac{V_0}{U_0}$ and $\theta = -\tan^{-1}\left(\frac{V_0}{U_0}\right)$. We set $V_0 = C$, $U_0 = 1$ and obtain $\tan^{-1} C = -\theta$. In summary, the iterative formulas to be calculated are

$$
\begin{array}{llll}
(i) & U_{i+1} = U_i - s_i 2^{-i} \cdot V_i \, ; & U_0 = 1, & \\
(i)' & V_{i+1} = V_i + s_i 2^{-i} \cdot U_i \, ; & V_0 = C, & (9.42) \\
(ii) & y_{i+1} = y_i - s_i \tan^{-1}(2^{-i}) \, ; & y_0 = 0, &
\end{array}
$$

with $s_i \in \{-1, 1\}$ set according to the signs of V_i and U_i so that V_{i+1} is closer to 0.

Example 9.4

To calculate $\tan^{-1}(1.0)$ in 10-bit precision we use the entries of Table 9.2. The steps of the calculation are shown in the table below. The final result is $y_{11} = 0.1100100100_2$ which is equal to the "exact" result in 10-bit precision.

i	U_i	V_i	y_i	s_i
1	1.0000000000	1.0000000000	0.0000000000	-1
2	1.1000000000	0.1000000000	0.0111011011	-1
3	1.1010000000	0.0010000000	0.1011010110	-1
4	1.1010010000	-0.0001010000	0.1101010101	1
5	1.1010010101	0.0000011001	0.1100010101	-1
6	1.1010010110	-0.0000011100	0.1100110101	1
7	1.1010010110	-0.0000000010	0.1100100101	1
8	1.1010010110	0.0000001011	0.1100011101	-1
9	1.1010010110	0.0000000100	0.1100100001	-1
10	1.1010010110	0.0000000001	0.1100100011	-1
11	1.1010010110	-0.0000000001	0.1100100100	

$\square$

9.5 THE HYPERBOLIC FUNCTIONS

To calculate the hyperbolic functions

$$\sinh x_0 = \frac{1}{2}\left(e^{x_0} - e^{-x_0}\right) \quad \text{and} \quad \cosh x_0 = \frac{1}{2}\left(e^{x_0} + e^{-x_0}\right), \tag{9.43}$$

we use an algorithm similar to that used for the exponential function; i.e.,

$$\begin{array}{rlll} (i) & x_{i+1} & = & x_i - g(b_i)\,, \\ (ii) & y_{i+1} & = & y_i \cdot b_i\,, & b_i = 1 + s_i 2^{-i}. \end{array} \tag{9.44}$$

We first rewrite b_i as

$$1 + s_i 2^{-i} = \sqrt{1 - (s_i 2^{-i})^2} \cdot \exp(\tanh^{-1}(s_i 2^{-i})) \tag{9.45}$$

which is based on the identity

$$1 + x = \sqrt{1 - x^2} \cdot \exp(\tanh^{-1} x) \quad \text{if } |x| \neq 1. \tag{9.46}$$

The proof of this identity is left to the reader as an exercise.

Formula (ii) in Equation (9.44) results in

$$y_m = y_0 \prod_{l=0}^{m-1} b_l = y_0 \left(\prod_{l=0}^{m-1} \sqrt{1 - (s_l 2^{-l})^2} \right) \cdot \exp\left(\sum_{l=0}^{m-1} \tanh^{-1}(s_l 2^{-l}) \right)$$

$$= y_0 \cdot \hat{K} \cdot \exp\left(\sum_{l=0}^{m-1} \tanh^{-1}(s_l 2^{-l}) \right) \tag{9.47}$$

where

$$\hat{K} = \prod_{l=0}^{m-1} \sqrt{1 - (s_l 2^{-l})^2}.$$

This is a constant factor ($\hat{K} = 1.205$) if we restrict s_l to $\{-1, 1\}$, as we did for the trigonometric functions.

Unlike the algorithm for e^x, we select here $g(b_i) = \tanh^{-1}(s_i 2^{-i})$, so

$$\sum_{l=0}^{i} \tanh^{-1}(s_l 2^{-l}) \to x_0 \quad \text{when} \quad x_{i+1} \to 0.$$

Thus, from Equation (9.47), $y_m = y_0 \cdot \hat{K} \cdot e^{x_0}$. We now define a new variable t_i, which is calculated recursively through the equation $t_{i+1} = t_i \cdot (1 - s_i 2^{-i})$. As in

Equation (9.47), this new variable converges to $t_0 \cdot \hat{K} \cdot e^{-x_0}$. Let $Z_i = \frac{1}{2} \cdot (y_i + t_i)$ and $W_i = \frac{1}{2} \cdot (y_i - t_i)$. Formula (ii) is replaced by the following two formulas:

$$\begin{array}{rll} (ii) & Z_{i+1} & = Z_i + s_i 2^{-i} \cdot W_i, \\ (ii)' & W_{i+1} & = W_i + s_i 2^{-i} \cdot Z_i. \end{array} \tag{9.48}$$

These variables converge to

$$\begin{aligned} Z_m & = \frac{1}{2} y_0 \hat{K} e^{x_0} + \frac{1}{2} t_0 \hat{K} e^{-x_0} \\ & = \frac{1}{2}(y_0 + t_0) \cdot \hat{K} \cdot \frac{e^{x_0} + e^{-x_0}}{2} + \frac{1}{2}(y_0 - t_0) \cdot \hat{K} \cdot \frac{e^{x_0} - e^{-x_0}}{2} \\ & = Z_0 \cdot \hat{K} \cdot \cosh x_0 + W_0 \cdot \hat{K} \cdot \sinh x_0 \end{aligned} \tag{9.49}$$

and

$$\begin{aligned} W_m & = \frac{1}{2} y_0 \hat{K} e^{x_0} - \frac{1}{2} t_0 \hat{K} e^{-x_0} \\ & = Z_0 \cdot \hat{K} \cdot \sinh x_0 + W_0 \cdot \hat{K} \cdot \cosh x_0 \end{aligned} \tag{9.50}$$

Now, setting $W_0 = 0$ and $Z_0 = 1/\hat{K}$ yields $Z_m = \cosh x_0$ and $W_m = \sinh x_0$.

The resulting formulas (i), (ii), and $(ii)'$ are similar to those for the trigonometric functions. There is, however, a major difference between the convergence of x_{i+1} to 0 in these two schemes. For the trigonometric functions the relationship

$$\tan^{-1}(2^{-(i+1)}) > 1/2 \cdot \tan^{-1}(2^{-i})$$

holds. As a result, even if we obtain $x_i = 0$ and $|x_{i+1}|$ becomes $\tan^{-1}(2^{-i})$ (since s_i is restricted to $\{-1, 1\}$), we can still expect x_m to converge to 0. However, for hyperbolic functions we have

$$\tanh^{-1}(2^{-(i+1)}) < 1/2 \cdot \tanh^{-1}(2^{-i})$$

and the convergence of x_m to 0 is not guaranteed unless several steps are repeated twice. For example, if $n = 24$ then steps 3, 4, 7, 12, 13, 18, and 21 must be repeated twice [8].

9.6 BOUNDS ON THE APPROXIMATION ERROR

In this section we estimate the maximum error that is to be expected when evaluating elementary functions using either additive normalization or multiplicative normalization.

In the additive normalization procedure, if x_0 is an n-bit fraction, then $\sum_{l=0}^{m-1} g(b_l)$ approaches x_0 with an error of $\epsilon = x_0 - \sum_{l=0}^{m-1} g(b_l)$, which satisfies $|\epsilon| \leq 2^{-n}$. At the same time, we attempt to evaluate $y = F(x_0)$ by calculating

$$y = F(\sum_{l=0}^{m-1} g(b_l)) = F(x_0 - \epsilon).$$

The corresponding Taylor series expansion is

$$y = F(x_0 - \epsilon) = F(x_0) - \epsilon \cdot \frac{dF}{dx}\big|_{x_0} + \frac{\epsilon^2}{2} \cdot \frac{d^2 F}{dx^2}\big|_{x_0} + 0(\epsilon^3). \qquad (9.51)$$

Since ϵ is of the order of 2^{-n}, the last two terms in the above expansion are negligible. As a result, the error δ in y is of the order of $\epsilon \cdot \frac{dF}{dx}\big|_{x_0}$. The magnitude of this error depends on the specific function $F(x_0)$. For example, for the exponential function $F(x_0) = e^{x_0}$, $\frac{dF}{dx}\big|_{x_0} = e^{x_0}$, and thus $|\delta| \leq 2^{-n} \cdot e^{\ln 2} = 2^{-(n-1)}$ for $x \leq \ln 2$. The maximum error in the function has double the size of the error in the argument. This error can be reduced by increasing the number of bits in x_i by 1.

In the multiplicative normalization procedure,

$$x_0 \prod_{l=0}^{m-1} b_l \to 1,$$

and the error is

$$\epsilon = 1 - x_0 \prod_{l=0}^{m-1} b_l \quad \text{or} \quad x_0 \prod_{l=0}^{m-1} b_l = 1 - \epsilon.$$

Instead of $F(x_0)$, we calculate

$$y = F\left(\frac{1}{\prod_{l=0}^{m-1} b_l}\right) = F\left(\frac{x_0}{1-\epsilon}\right) \approx F(x_0 \left[1 + \epsilon - \epsilon^2 + \cdots\right]) \approx F(x_0 + x_0\epsilon).$$

The Taylor series expansion of the last expression is

$$F(x_0 + x_0\epsilon) = F(x_0) + x_0\epsilon \cdot \frac{dF}{dx}\big|_{x_0} + \frac{1}{2}(x_0\epsilon)^2 \cdot \frac{d^2 F}{dx^2}\big|_{x_0} + 0(\epsilon^3). \qquad (9.52)$$

For example, for the natural logarithm function $F(x_0) = \ln x_0$, the derivatives are $\frac{dF}{dx}\big|_{x_0} = \frac{1}{x_0}$ and $\frac{d^2 F}{dx^2}\big|_{x_0} = -\frac{1}{x_0^2}$. Therefore, $y = F(x_0) + \epsilon + 0(\epsilon^2)$. Hence, the error δ in y satisfies $|\delta| \approx |\epsilon| \leq 2^{-n}$, and the precision obtained is satisfactory.

9.7 SPEED-UP TECHNIQUES

Consider the exponential function $F(x_0) = e^{x_0}$, for which the precalculated constants $\ln(1 + s_i 2^{-i})$ have the following Taylor series expansion:

$$\ln(1 + s_i 2^{-i}) = s_i 2^{-i} - (s_i)^2 2^{-2i-1} + \frac{1}{3} s_i 2^{-3i} - \cdots$$

For $i > n/2$, the above expression is approximately $s_i 2^{-i}$. As a result, not only can we reduce the size of the ROM required but, more importantly, the last $n/2$ steps can be replaced by a single operation, as shown below. In the steps prior to step i, we have already canceled at least the first $(i-1)$ bits in x_i. x_i thus has the following form:

$$x_i = 0.\underbrace{0 \cdots 00}_{i-1} z_i z_{i+1} \cdots$$

where z_k $(k = i, i+1, \cdots)$ is a single bit. To cancel the remaining nonzero bits in x_i, for $i \geq n/2$, we should select $s_k = z_k$ $(k > i)$; i.e., all the remaining s_k's can be predicted ahead of time. Based on this knowledge, we may speed up the execution. In the last steps we need to calculate (for $i \geq n/2$)

$$y_m = y_i \prod_{k=i}^{m-1} (1 + s_k 2^{-k}) \approx y_i \left(1 + s_i 2^{-i} + s_{i+1} 2^{-(i+1)} + s_{i+1} 2^{-(i+1)} + \cdots \right),$$

$$(9.53)$$

and, since we select $s_k = z_k$ for $k \geq i$, the last term in Equation (9.53) equals $(1 + x_i)$. Thus,

$$y_m = y_i(1 + x_i). \tag{9.54}$$

If a fast multiplier is available, the overall execution time can be reduced. This is called a *termination algorithm* [2] (or terminal linear approximation). Even if a fast multiplier is not available, we can still take advantage of the ability to predict the s_i's by performing all the additions of the products in Equation (9.53) in a carry-save manner, avoiding the time-consuming carry-propagation.

We may arrive at the same expression for the terminal approximation in a different way based on the Taylor series expansion

$$\begin{aligned} F(x_0) &= F((x_0 - x_i) + x_i) \\ &= F(x_0 - x_i) + x_i \cdot \frac{dF}{dx}\Big|_{(x_0-x_i)} + \frac{1}{2} x_i^2 \cdot \frac{d^2 F}{dx^2}\Big|_{(x_0-x_i)} + 0(x_i^3) \end{aligned}$$

$$(9.55)$$

for $|x_i| \leq 2^{-n/2}$. It can be shown that $F(x_0 - x_i) = \frac{dF}{dx}\big|_{(x_0-x_i)} = \frac{d^2 F}{dx^2}\big|_{(x_0-x_i)} = y_i$, and therefore, $F(x_0) = y_i(1 + x_i) + \delta$, where $|\delta| \approx \frac{1}{2} x_i^2 \cdot y_i \leq \frac{1}{2} \cdot 2^{-n} \cdot e^{\ln 2} = 2^{-n}$.

Thus, the bound on the error when the terminal approximation is used is half its value without the terminal approximation, providing higher precision.

A termination algorithm can also be applied in the calculation of the natural logarithm function, $F(x_0) = \ln x_0$. In this case, x_i has the form

$$1 \; - \; x_i = 0.\underbrace{0 \, \cdots \, 00}_{i-1} \; z_i z_{i+1} \cdots$$

or

$$x_i = 1 - z_i 2^{-i} - z_{i+1} 2^{-(i+1)} - \cdots = 1 - \sum_{k=i}^{m-1} z_k 2^{-k}.$$

The formula for x_{i+1} yields

$$x_{i+1} = x_i \cdot (1 + s_i 2^{-i}) = 1 + s_i 2^{-i} - z_i 2^{-i} - \cdots$$

For x_{i+1} to approach 1, we should therefore select $s_k = z_k$ for $k \geq i$. In parallel, we expect to calculate in the remaining steps

$$y_m \; = \; y_i \; - \; \sum_{k=i}^{m-1} \ln(1 \; + \; s_k 2^{-k}).$$

Consequently, based on the Taylor series expansion for $\ln(1 + s_k 2^{-k})$, we obtain the following terminal approximation for $i \geq n/2$:

$$y_m \approx y_i - \sum_{k=i}^{m-1} s_k 2^{-k} = y_i - (1 - x_i) \tag{9.56}$$

The evaluation of the trigonometric functions can also be accelerated by predicting the s_i's based on the series expansion in Equation (9.33). Previously, in order to obtain a constant value for K independent of the selected s_i's, we have restricted s_i to be in $\{-1, 1\}$. However, if

$$K = \prod_{l=0}^{m-1} \sqrt{1 + 2^{-2l}}$$

is replaced by

$$K = \prod_{l=0}^{n/2} \sqrt{1 + 2^{-2l}},$$

the deviation from the exact value is less than 2^{-n} [1]. Therefore, for $i > n/2$, we may select s_k $(k \geq i)$ from the set $\{0, 1\}$ and predict $s_k = z_k$, where the z_k are the bits in the least significant portion of x_i:

$$x_i = 0.\underbrace{0 \, \cdots \, 00}_{i-1} \; z_i z_{i+1} \cdots$$

Being able to predict the s_k's allows us to perform the additions in the equations for x_i, Z_i and W_i in a carry-save manner [1]. Reexamining the series expansion of $\tan^{-1}(1 + s_l 2^{-l})$ in Equation (9.33), we conclude that we may, in principle, start the prediction process even earlier and predict the s_l's for $i \leq l \leq 3i$ all at once. The corresponding terms can then be added with a carry-save adder and a single pass through a carry-propagate adder. However, for $l < n/2$ we still need to use the set $\{-1, 1\}$. This can be done by properly recoding the bits in x_i using the set $\{-1, 1\}$.

We would also like to speed up the computation for the first steps, $0 \leq i \leq n/2$. This can be done by using a radix higher than 2; e.g., $g(b_i) = 1 + s_i r^{-i}$, where r is some power of 2, say $r = 2^q$, and $s_i \in \{-(r-1), -(r-2), \cdots -1, 0, 1, \cdots, (r-1)\}$. This way, we may handle q bits of x in a single step, but with an increased complexity of each step, since s_i has a larger range. Still, some improvement in speed is possible [4].

Another method is to allow s_i to assume the values in $\{-1, 0, 1\}$ and modify the selection rule in such a way that the probability of selecting $s_i = 0$ is maximized. This is similar to the idea behind the variations of the *SRT* division algorithm. This approach has been analyzed in [3] but has not gained much popularity.

9.8 OTHER TECHNIQUES FOR EVALUATING ELEMENTARY FUNCTIONS

Many calculators and floating-point processors employ some variations of the previously presented algorithms for the evaluation of elementary functions. Usually, each one of these has a particular implementation that depends on the precision and speed requirements and the area constraints. Still, several other methods for calculating elementary functions have been proposed, and some have been implemented; e.g., [5]. In [9] an evaluation of elementary functions based on rational approximations is proposed. This method, which is commonly used when evaluating elementary functions in software, can become very cost-effective for hardware implementation. This is the case when a fast adder and multiplier are available and when high precision is required; e.g., the arguments are extended double-precision floating-point numbers in the IEEE standard. In such a situation, hardware implementation of rational approximations can successfully compete with the methods based on continued summations and multiplications (whose convergence is linear in the number of bits) when execution time and chip area are considered.

A somewhat different approach combines polynomial approximations with a look-up table [10, 16]. Here, the domain of the argument x of the elementary function $f(x)$ is divided into smaller intervals (usually of equal size) and the values of $f(x_i)$ for the boundary points x_i between the intervals are kept in the look-up table. Then, the value of $f(x)$ at the given point x is calculated based on the value $f(x_i)$, which is read from the table where x_i is the closest boundary point to x, and a polynomial approximation $p(x - x_i)$ for $f(x - x_i)$. Since the distance $(x - x_i)$ is small, a very simple polynomial can be employed, requiring significantly less time than a rational approximation for $f(x)$ on the entire domain. The overall algorithm therefore has three steps:

1. Find the closest boundary point x_i and calculate the "reduction transformation," which is usually the distance $d = x - x_i$.

2. Calculate $p(d)$.

3. Combine $f(x_i)$ with $p(d)$ to calculate $f(x)$.

Example 9.5

The following algorithm can be employed for calculating 2^x on $[-1, 1]$ [16]: 32 boundary points of the form $x_i = i/32$, $i = 0, 1, \cdots 31$, are defined, and the values of 2^{x_i} are precalculated and stored in a look-up table. In step 1, we search for an x_i such that $|x - m - x_i| \leq 1/64$, where $m = -1, 0$, or 1. Then, we calculate $d = (x - m - x_i) \cdot \ln 2$. This "distance" satisfies $e^d = 2^{x-m-x_i}$, and $|d| \leq (\ln 2)/64$. In step 2 we calculate an approximation for $e^d - 1$ using a polynomial $p(d) = d + p_2 d^2 + p_3 d^3 + \cdots p_k d^k$, where p_2, $p_3 \cdots p_k$ are precalculated coefficients of the polynomial approximation of the function $e^d - 1$ on the interval $[-(\ln 2)/64, (\ln 2)/64]$. In step 3 we reconstruct 2^x using

$$2^x = 2^{m+x_i} \cdot e^d = 2^m \left(2^{x_i} + 2^{x_i} \cdot (e^d - 1)\right) \approx 2^m \left(2^{x_i} + 2^{x_i} \cdot p(d)\right)$$

where 2^{x_i} is read out of the look-up table. A detailed error analysis for this algorithm for IEEE double-precision arguments shows that the error is bounded by 0.556 ulp, where ulp for this format equals 2^{-52}.   □

9.9 EXERCISES

9.1. Apply the procedure in Section 9.1 to calculate $e^{0.5}$. Assume that the argument $x_0 = 0.5$ and all intermediate results have 12 fractional bits. Prepare a table of all terms of the form $\ln(1 \pm 2^{-i})$ with 12-bit precision. Compare your result to the exact value of $\sqrt{e}$ and compare the error to 2^{-12}.

9.2. Repeat problem 1, applying the procedure in Section 9.2 for calculating $\ln 0.5$.

9.3. Prove identity (9.46).

9.4. Prove that Equation (9.55) yields the same expression for the termination algorithm (for e^x) as Equation (9.54).

9.5. Write a procedure for calculating y_0/x_0 using multiplicative normalization. Devise a termination algorithm and discuss its effectiveness.

9.6. The reciprocal of the square root of a given n-bit operand can be calculated in n steps using the following two equations:

$$(i) \quad x_{i+1} = x_i \cdot (1 + s_i 2^{-i})^2 = x_i \cdot (1 + 2s_i 2^{-i} + s_i^2 2^{-2i})$$

$$(ii) \quad y_{i+1} = y_i \cdot (1 + s_i 2^{-i})$$

where the s_i's are selected so that $x_{i+1} \to 1$ and consequently $y_{i+1} \to y_0/\sqrt{x_0}$. We want to examine the possibility of employing a termination algorithm in the kth step, in order to speed up the computation by replacing the remaining $(n-k)$ steps by a single operation. In step k we have

$$1 - x_k = 1 - x_0 \prod_{l=0}^{k-1} (1 + s_l 2^{-l})^2 = 0.0 \cdots 0 z_k z_{k+1} \cdots$$

and to obtain the final value of y, we calculate $y_k \prod_{l=k}^{n} (1 + s_l 2^{-l})$.

For what values of k can the termination algorithm be used? Write in detail the computation needed in this termination algorithm.

9.7. Verify that if the factor $K = \prod_{l=0}^{n} \sqrt{1 + 2^{-2l}}$ in the calculation of sine/cosine is

replaced by $K = \prod_{l=0}^{n/2} \sqrt{1 + 2^{-2l}}$, then the deviation from the exact value is less

than 2^{-n}.

9.8. Find the reduction procedure that should be used when calculating $\sin(x)$ for $x \geq \pi/2$ using the algorithm in Section 9.3.

9.9. Show that selecting $g(b_i) = b_i^k$, as in Equation (9.18), yields

$$\frac{y_{i+1}}{x_{i+1}^k} = \frac{y_i}{x_i^k}$$

9.10. After the first step in Example 9.3 we already obtain $x = 0$. Why do we have to execute the remaining nine steps?

9.10 REFERENCES

[1] P. W. BAKER, "Suggestion for a fast binary sine/cosine generator," *IEEE Trans. on Computers, C-25* (Nov. 1976), 1134-1137.

[2] T. C. CHEN, "Automatic computation of exponentials, logarithms, ratios, and square roots," *IBM Journal Res. and Dev., 16* (July 1972), 380-388.

[3] B. D. DELUGISH, "A class of algorithms for automatic evaluation of certain elementary functions," Dept. Comp. Sci., Univ. of Illinois, Rep. 399, June 1970.

[4] M. D. ERCEGOVAC, "Radix-16 evaluation of certain elementary functions," *IEEE Trans. on Computers, C-22* (June 1973), 561-566.

[5] P. M. FARMWALD, "High bandwidth evaluation of elementary functions," *Proc. of 5th Symp. on Computer Arithmetic* (May 1981), 139-142.

[6] D. GOLDBERG, private communication.

[7] J. F. HART *et al.*, *Computer approximations*, Wiley, New York, 1968.

[8] G. L. HAVILAND and A. A. TUSZYNSKI, "A CORDIC arithmetic processor chip," *IEEE Trans. on Computers, C-29* (Feb. 1980), 68-79.

[9] I. KOREN and O. ZINATY, "Evaluating elementary functions in a numerical co-processor based on rational approximations," *IEEE Trans. on Computers, 39* (Aug. 1990), 1030-1037.

[10] P. W. MARKSTEIN, "Computation of elementary functions on the IBM RISC system/6000 processor," *IBM Journal Res. and Dev., 34* (Jan. 1990), 111-119.

[11] J.-M. MULLER, *Elementary functions: Algorithms and implementation*, Birkhauser, 1997.

[12] R. NAVE, "Implementation of transcendental functions on a numeric processor," *Microprocessing and Microprogramming, 11* (1983), 221-225.

[13] M. J. SCHULTE and E. E. SWARTZLANDER, "Hardware designs for exactly rounded elementary functions, *IEEE Trans. on Computers, 43* (August 1994), 964-973.

[14] W. H. SPECKER, "A class of algorithms for ln x, exp x, sin x, cos x, $\tan^{-1} x$, $\cot^{-1} x$," *IEEE Trans. on Electron. Computers, EC-14* (Feb. 1965), 85-86.

[15] S. STORY and P. T. P. TANG, "New algorithm for improved transcendental functions on IA-64," *Proc. of 14th Symp. on Computer Arithmetic* (April 1999), 4-11.

[16] P. T. P. TANG, "Table-lookup algorithms for elementary functions and their error analysis," *Proc. of 10th Symp. on Computer Arithmetic*, (1991), 232-236.

[17] J. E. VOLDER, "The CORDIC trigonometric computing technique," *IRE Trans. on Electron. Computers, EC-8* (Sept. 1959), 330-334.

[18] J. S. WALTHER, "A unified algorithm for elementary functions," *Spring Joint Computer Conf., Proc., 38* (1971), 379-385.

10

LOGARITHMIC NUMBER SYSTEMS

A number system based on logarithms can simplify multiplication, division, roots, and powers. When logarithms are used, multiplication and division are reduced to addition and subtraction, respectively, and powers and roots are reduced to multiplication and division, respectively. On the other hand, add and subtract operations become more complex. Another major problem is deriving logarithms and antilogarithms quickly and accurately enough to allow conversions to and from the conventional number representations. These conversions always involve approximations, resulting in inaccuracies. Therefore, binary logarithms can be useful only in arithmetic units dedicated to special applications where very few conversions are required but many multiplications and divisions are executed; e.g., real-time digital filters.

10.1 SIGN-LOGARITHM NUMBER SYSTEMS

Let a number A be represented by a sign digit S_A and a logarithm E_A that includes an integer part and a fractional part

$$S_A E_A \;=\; S_A \underbrace{a_{k-1} a_{k-2} \cdots a_1 a_0}_{k} \cdot \underbrace{a_{-1} a_{-2} \cdots a_{-l}}_{l} \qquad (10.1)$$

requiring a total of $n = 1 + k + l$ bits. The sign S_A is set to 0 if A is positive, and to 1 if A is negative. E_A is the logarithm of the absolute value of A; i.e., $E_A = \log_2 |A|$. The interpretation rule for $S_A E_A$ is thus

$$A = (-1)^{S_A} \cdot 2^{E_A}. \qquad (10.2)$$

The base of the exponent may, in general, be different from 2.

247

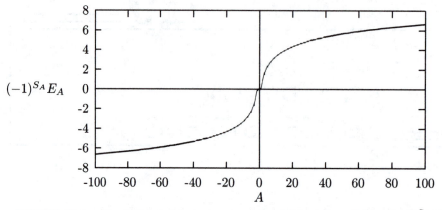

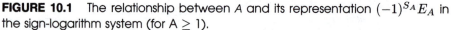

FIGURE 10.1   The relationship between A and its representation $(-1)^{S_A} E_A$ in the sign-logarithm system (for $A \geq 1$).

To represent numbers smaller than 1, negative logarithms are needed. For this purpose, we may use the two's complement representation or a biased representation, which takes the general form

$$E_A = (a_{k-1} \cdots a_0 \cdot a_{-1} \cdots a_{-l})_2 - Bias.$$

Commonly used values for the bias are 2^{k-1} or $2^{k-1} - 1$.

Examining Equation (10.2), one should realize that this is an extreme case of the familiar floating-point system with the significand always equal to 1. As a result, the exponent E_A is allowed to be a mixed number rather than an integer, in order to enable the representation of numbers that are not integral powers of the base. In this number system, as in the floating-point system, zero is not included in the ordinary range of values. If biased logarithms are used, we may decide that $E_A = 0$ represents 0 instead of $2^{E_{min}}$. Another way to represent zero is to have a special bit in the format indicating that the value is 0, regardless of the value of E_A [8].

Figure 10.1 depicts the relationship, for $A \geq 1$, between the real number A and its representation $(-1)^{S_A} E_A$ in the sign-logarithm system. It is evident from this figure that $(-1)^{S_A} E_A$ is monotonic in A, and comparison is therefore straightforward. Given two numbers $A = (-1)^{S_A} \cdot 2^{E_A}$ and $B = (-1)^{S_B} \cdot 2^{E_B}$, we first compare their signs S_A and S_B; then, if the signs are equal, we compare their logarithms E_A and E_B.

Example 10.1

Let k be 4 and l equal 3. Consequently, n equals 8. Suppose that the logarithm E_A is represented in the two's complement method. The following are two representations within the range:

00001.010   represents the numerical value $+2^{(1+\frac{1}{4})}$ $=$ $+2.37841_{10}$

01110.100   represents the numerical value $+2^{-(1+\frac{1}{2})}$ $=$ $+0.35355_{10}$

The largest positive number is $00111.111 = 2^{(8-\frac{1}{8})} = 234.7530_{10}$. The smallest positive number is $01000.000 = 2^{-8} = 0.003906_{10}$. There is no representation for zero.   □

10.2 ARITHMETIC OPERATIONS

In a logarithmic system multiplication and division are simpler operations than addition and subtraction. To calculate the product P of two operands A and B we add their logarithms

$$E_P = E_A + E_B, \tag{10.3}$$

and if these logarithms are biased, we have to then subtract the bias. As with conventional signed-magnitude systems, the sign bit of the product is determined by the modulo 2 addition of the operands' sign bits, $S_P = S_A \oplus S_B$. Similarly, when calculating the quotient $Q = A/B$ we subtract the logarithms

$$E_Q = E_A - E_B, \tag{10.4}$$

and if these logarithms are biased, we must add the bias. The sign bit of the quotient, S_Q, is determined in exactly the same way as is done for the product; i.e., $S_Q = S_A \oplus S_B$. Comparing the multiply and divide operations to their counterparts in a floating-point system, one should note that in the logarithmic system they are exact operations, and no rounding is required. Thus, these two operations do not contribute to the accumulation of computation errors. Overflow and underflow may occur when executing the add or subtract operations in Equations (10.3) and (10.4), but these are easily detected.

In contrast, addition and subtraction of operands that are given by their logarithms are complicated. One brute-force way to perform these operations is through the use of a complete look-up table. However, the size of such a table ($2^{2n} \times n$ bits) is prohibitive for any reasonable value of n, the number of bits in an operand. We may therefore use one of the following two alternatives, which still require tables, but of smaller size:

1. Use an antilogarithm table, add and then use a logarithm table. This requires a total of $3 \cdot 2^n \times n$ bits in tables (three tables if the antilogarithms of the two operands are to be read simultaneously).

2. Calculate directly the approximated sum (or difference). This method employs smaller look-up tables and is therefore the one most commonly used.

When calculating the sum (or difference)

$$C = A \pm B = (-1)^{S_C} \cdot 2^{E_C}$$

according to the second method, we distinguish between two cases. In the first, $|A| > |B|$, and we rewrite the expression for C as

$$C = A \pm B = A(1 \pm \frac{B}{A}).$$

We set S_C equal to S_A and

$$E_C = \log_2 |A(1 \pm \frac{B}{A})| = \log_2 |A| + \log_2 |1 \pm \frac{B}{A}| = E_A + \Phi\left(E_A - E_B\right), \quad (10.5)$$

where the function $\Phi\left(E_A - E_B\right)$ is defined as

$$\Phi\left(E_A - E_B\right) = \log_2 |1 \pm \frac{B}{A}| = \log_2 |1 \pm 2^{-(E_A - E_B)}|. \qquad (10.6)$$

The value of $\Phi(x)$, where $x = E_A - E_B > 0$, must be precalculated and stored in a table. For convenience, two separate tables are commonly used, one for $\Phi^+(x) = \log_2(1 + 2^{-x})$ and a second for $\Phi^-(x) = \log_2(1 - 2^{-x})$. Each table can be implemented in a ROM of a size not larger than $2^n \times n$. The size of the table for $\Phi^+(x)$ may be reduced to $2^n \times l$, since $\Phi^+(x) \leq 1$ for $x \geq 0$. In other words, $\Phi^+(x)$ is always a fraction and will never require more than l bits.

In the second case, $|A| < |B|$, so $S_C = S_B$ and $E_C = E_B + \Phi(E_B - E_A)$. Consequently, the steps in both cases are:

1. Compare A and B to determine the larger of the two.
2. Calculate $x = E_A - E_B$ or $E_B - E_A$.
3. Read $\Phi^+(x)$ or $\Phi^-(x)$ from the appropriate table and add it to either E_A or E_B.

The first two steps can be executed in parallel if the data flow depicted in Figure 10.2 is adopted [7]. As a result, the total time needed is approximately

$$T_{ADD/SUB} \approx 2 \cdot T_{ADDER} + T_{ROM}. \qquad (10.7)$$

The only source of error when performing an addition (or subtraction) is the rounding of Φ^+ or Φ^-. The values stored in the ROM should be rounded (e.g., to nearest, see Chapter 4) rather than truncated and the error introduced will therefore be no larger than $\frac{1}{2} \cdot 2^{-l}$.

The size of the above two tables for Φ^+ and Φ^- is the major obstacle when attempting to implement an arithmetic unit that operates on sign-logarithm numbers. For $n \geq 20$, the required ROM becomes prohibitively large. Therefore,

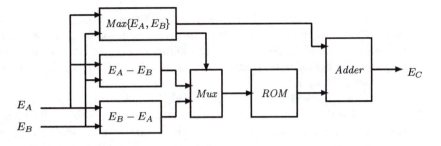

FIGURE 10.2   Adder/subtractor for sign-logarithm numbers.

several strategies for reducing the size of the look-up tables have been suggested and implemented. One approach is to partition the table of size $2^n \times n$ into several smaller tables [7]. Since $\Phi(x)$ decreases rapidly with increasing x, the size of the corresponding partial tables can be substantially reduced. Another approach is to use a combination of linear approximation and a look-up table [8]. In either case, there is no need to generate $\Phi(x)$ for very large values of x, since the value of $\Phi(x)$ becomes smaller than the resolution of the system.

Example 10.2

A 20-bit logarithmic number system processor has been designed and implemented as a single VLSI chip [7]. The 20 bits include a sign bit and 19 bits for an exponent in two's complement representation. These 19 bits are partitioned into a sign bit, a 6-bit integer part, and a 12-bit fractional part. The two look-up tables for Φ^+ and Φ^-, if fully implemented in ROM, would require $2^{18} \cdot 12 + 2^{18} \cdot 18 = 7.8\text{Mbits}$. The size of these tables can be substantially reduced as described next. First, there is no need to generate values that are smaller than 2^{-12}, the resolution of the selected number system. The solution of

$$\Phi^+(x) = \log_2(1 + 2^{-x}) = 2^{-12}$$

is $x = 12.52$ and, the solution of $\Phi^-(x) = \log_2(1 - 2^{-x}) = -2^{-12}$ is also $x = 12.52$. Thus, no look-up table entries are required for $x \geq 12.52$. Consequently, there is no need to provide more than 4 bits in the integer part of x as inputs to the look-up tables, allowing $x \leq 15$. These four bits, together with the 12 fractional bits, constitute the required 16 inputs to the ROM instead of 18. The remaining range $[0, 15]$ is then divided into 11 smaller intervals: $[0.0, 0.5]$, $[0.5, 1.0]$, $[1.0, 2.0]$, $[2.0, 3.0]$, $[3.0, 4.0]$, $[4.0, 5.0]$, $[5.0, 6.0]$, $[6.0, 7.0]$, $[7.0, 8.0]$, $[8.0, 9.0]$, and $[9.0, 15.0]$. For the first 10 intervals ROM1 through ROM10 are used while for the last one a PLA implementation has proven to be more economical, providing values for the subinterval $[9, 12.52]$. The reason behind the partition of the interval $[0.0, 9.0]$ into 10 smaller intervals is that the graph for $\Phi(x)$ becomes flat for large values of x. As a result, the number of input bits to the ROM and the number of output bits decrease rapidly. The total ROM space employed is 83.55 Kbits.

This approach, if applied to a 30-bit format, would require a ROM space of 70Mbits [8]. Therefore, a different approach, using a linear approximation, was proposed to further reduce the size of the look-up tables. With this approach, the size of the ROM decreases. However, the execution time increases. □

An important advantage of the logarithmic system is that several other arithmetic operations can be executed in a straightforward manner. For example, the reciprocal of a given number $A = (-1)^{S_A} 2^{E_A}$ is simply $A^{-1} = (-1)^{S_A} 2^{-E_A}$, and only the two's complement of E_A needs to be calculated. Squaring a given number is accomplished by $A^2 = 2^{2E_A}$, requiring a shift left operation. If the logarithm is biased, the bias must be subtracted. The square root of A is given by $\sqrt{A} = 2^{E_A/2}$, requiring a shift right operation. Here, if the logarithm is biased, the bias must be added first. Exponentiation is also simplified, since $A^y = 2^{y \cdot E_A}$, and only a fixed-point multiplication is required.

10.3 COMPARISON TO BINARY FLOATING-POINT NUMBERS

As previously noted, the logarithmic system is an extreme case of the conventional floating-point system. Therefore, a more detailed comparison between the two should be made. Two important characteristics that need to be analyzed are the range and the accuracy of representation.

The range of positive logarithms E_A using $k + l$ bits is

$$-2^{k-1} \le E_A \le 2^{k-1} - 2^{-l} \approx 2^{k-1}.$$

Therefore, the range for positive numbers in the sign-logarithm system is

$$2^{-2^{k-1}} \le A^+ \le 2^{2^{k-1}}. \tag{10.8}$$

This range should be compared to the range of positive normalized floating-point numbers with $\beta = 2$ and $n = e + m + 1$, where e and m are the number of bits in the exponent and the normalized fractional significand, respectively. The latter range is, as shown in Chapter 4,

$$\frac{1}{2} \cdot 2^{-2^{(e-1)}} \le F^+ \le (1 - 2^{-m}) \cdot 2^{2^{(e-1)}-1}. \tag{10.9}$$

If we set $e = k$ and, consequently, $m = l$, the ranges remain about the same.

To measure the accuracy of representation in the logarithmic system, we calculate the relative step size, defined as

$$\frac{2^{(E_A + 2^{-l})} - 2^{E_A}}{2^{E_A}} = 2^{2^{-l}} - 1. \tag{10.10}$$

The maximum relative representation error, which equals half the relative step size, is thus a constant independent of E_A.

For the floating-point system the relative step size is

$$\frac{(M + 2^{-m})2^E - M \cdot 2^E}{M \cdot 2^E} = \frac{2^{-m}}{M}. \tag{10.11}$$

To compare the two step sizes, assume that $l = m$. Since

$$\lim_{l \to \infty} \frac{2^{2^{-l}} - 1}{2^{-l}} = 0.693,$$

and for normalized fractions $1 \leq 1/M \leq 2$, the following inequality holds:

$$2^{2^{-l}} - 1 \leq \frac{2^{-l}}{M}$$

As a result, the representation error in the logarithmic system is slightly lower than that in the corresponding floating-point system (with $m = l$), especially for small numbers. Numbers with the same exponent are equally spaced in the floating-point number system, while in the sign-logarithm system smaller numbers are denser.

Example 10.3

The 20-bit logarithmic processor reported in [7] has the range $2^{-64} \leq A^+ \leq 2^{64}$ with a precision of $2^{-12.52}$. The range of this system is twice that of a 20-bit floating-point format with a 12-bit significand and 7-bit exponent. The precision is slightly better, $2^{-12.52}$ versus 2^{-12}.   □

10.4 CONVERSIONS TO/FROM CONVENTIONAL REPRESENTATIONS

Conversions to the logarithmic number system from either a fixed-point system or a floating-point system require the calculation of logarithms. The opposite conversions require the calculation of antilogarithms. For example, to convert the floating-point number $(-1)^S \cdot M \cdot 2^E$ to the logarithmic system we must calculate $F = \log_2 M$ to obtain

$$(-1)^S \cdot M \cdot 2^E = (-1)^S \cdot 2^F \cdot 2^E = (-1)^S \cdot 2^{E+F}.$$

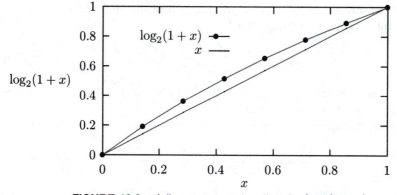

FIGURE 10.3   A linear approximation for $\log_2(1+x)$.

The logarithm of a given operand can be found in a look-up table whose size grows exponentially with the number of operand bits. Another way to find it is to calculate an approximation to the logarithm. Let N be a binary number $z_u z_{u-1} \cdots z_0 . z_{-1} \cdots z_{-v}$, and let z_t be the most significant nonzero bit of N. The value of this number can be written as

$$N = 2^t + \sum_{i=-v}^{t-1} 2^i z_i = 2^t\left(1 + \sum_{i=-v}^{t-1} 2^{i-t} z_i\right) = 2^t(1+x) \tag{10.12}$$

where x is a fraction, $0 \leq x < 1$. Clearly,

$$\log_2 N = t + \log_2(1+x) \tag{10.13}$$

where t is the characteristic of the logarithm and $\log_2(1+x)$ constitutes the mantissa.

A linear approximation for $\log_2(1+x)$, suggested in [5], uses only the linear term in the Taylor series; i.e., $\log_2(1 + x) \approx x$. This approximation is shown in Figure 10.3, and its error is $\epsilon(x) = \log_2(1+x) - x$. The maximum approximation error is found by differentiating $\epsilon(x)$, obtaining $Max\ \epsilon(x) = 0.08639$ for $x = 0.44269$.

The hardware implementation of this linear approximation is very simple. The operand N is stored in a shift register and a counter with an initial value of u is used. N is shifted to the left repeatedly until a 1 is shifted out, and at the same time the counter is decremented once for every shift operation. The contents of the counter at the end of the operation are $u - (u - t) = t$, and the contents of the shift register are the approximated mantissa x.

Example 10.4

Let $N = 0111.01 = 7.25_{10}$ and $u = 3$. We shift out a single zero, and then a 1. Hence, $t = 2$ and at the end the register contains 1101, interpreted as $.1101 = .8125_{10}$. Therefore, $\log_2 7.25 \approx 10.1101_2 = 2.8125_{10}$. The accurate value is 2.85798_{10}.

A similar approximation can be used for the antilogarithm. Given $t + y$, where y is a fraction, we need to calculate $N = 2^{t+y} = 2^t 2^y \approx 2^t(1 + y)$. This is the same approximation used before, $y \approx \log_2(1 + y)$. This can be implemented in hardware by placing a 1 in position t and placing the fraction y next to it. For example, given $100.1010 = 4.625_{10}$, $t = 4$ and the approximated antilogarithm is $11010.00_2 = 26_{10}$. The correct value is $2^{4.625} = 24.67537_{10}$. $\qquad\qquad\qquad\square$

A piecewise linear approximation has also been suggested [1]. The interval [0,1] is divided into four equal subintervals, and a linear approximation of the form $x + a \cdot f(x) + b$ is used for each subinterval, where $f(x)$ is either x or $\bar{x}$, the one's complement of x. The constants a and b are selected so as to minimize the error and be fractions with powers of 2 as denominators. The resulting expression is [1]

$$\log_2(1 + x) \approx \begin{cases} x + \frac{5}{16}\, x & 0 \le x < \frac{1}{4} \\ x + \frac{5}{64} & \frac{1}{4} \le x < \frac{1}{2} \\ x + \frac{1}{8}\bar{x} + \frac{3}{128} & \frac{1}{2} \le x < \frac{3}{4} \\ x + \frac{1}{4}\bar{x} & \frac{3}{4} \le x < 1 \end{cases} \qquad (10.14)$$

The approximation error is $-0.006 \le \epsilon \le 0.008$. The total error range, 0.014, is lower than that of the linear approximation by a factor of 6.

Higher precision can be achieved by using a look-up table implemented in a ROM. However, the size of this ROM is prohibitively large for reasonable values of the number of operand bits. A more economical implementation based on a PLA has been suggested in [4].

10.5 EXERCISES

10.1. (a) For a sign-logarithm system with $n = 16$, $k = 6$, and $l = 9$, find the smallest and largest positive numbers, assuming base 2 and that the two's complement method is used to represent negative logarithms. Calculate the maximum relative representation error.
(b) Repeat (a) for base 10.

10.2. Given the sign-logarithm system defined in problem 1, show the representation of the two operands $X = 2.5$ and $Y = 3.7$, in this system and perform the operations $X + Y$, $X - Y$, $X \cdot Y$, X/Y, $1/X$, X^2, $\sqrt{X}$, and X^Y. Calculate the entries of Φ^+ and Φ^- that are needed.

10.3. Write a Boolean expression for the signal, selecting the table of $\Phi^+(x)$ given the sign bits of the two operands, S_A and S_B, and the signals ADD and SUB(TRACT) indicating the type of operation being executed.

10.4. A 32-bit format for the sign-logarithm system has been suggested with $k = 8$, $l = 23$, a base 2, and a bias of 127 [3]. This results in a format that is very close to the single-precision format of the IEEE standard (see Chapter 4). Write down an expression for the value of a nonzero number X given the sign bit S_X and the logarithm E_X. Use the notation $E_X = I + F$, where I is the k-bit integer and F is the l-bit fraction. Compare the range of these two number representations and write the rule for converting an IEEE floating-point number to the sign-logarithm system. Estimate the conversion error and suggest a way to reduce it.

10.5. Determine the minimum number of inputs and outputs needed for ROM6, which corresponds to the range [4.0.5.0] of $\Phi^+(x)$ for the 20-bit logarithmic processor described in the text [7].

10.6. Show that the maximum error of the approximation $\log_2(1 + x) \approx x$ is 0.08639. Suppose that the approximation $\log_2(1 + x) \approx x + c_i$ is used instead, where the interval [0,1] is divided into four subintervals, as in Equation (10.14), and c_i is a constant employed for the ith subinterval ($i = 1, 2, 3, 4$). Find the best values for the c_i's so as to minimize the error, and calculate the resulting maximum error.

10.7. Write an expression for the distance between two adjacent numbers in the sign-logarithm system and compare it to that of the corresponding floating-point system. Show that smaller numbers are denser in the sign-logarithm system.

10.8. To represent values in the range $|A| \leq 1$ we may restrict $E_A = \log_b |A|$ to positive integers. What is the range of values that the base b may assume?

10.6 REFERENCES

[1] M. COMBET, H. V. ZONNEVELD, and L. VERBEEK, "Computation of the base two logarithm of binary numbers," *IEEE Trans. on Elect. Computers*, EC-14 (Dec. 1965), 863-867.

[2] A. D. EDGAR and S. C. LEE, "FOCUS microcomputer number system," *Communications of the ACM*, 22 (March 1979), 166-177.

[3] F. S. LAI and C. E. WU, "A hybrid number system processor with geometric and complex arithmetic capabilities," *IEEE Trans. on Computers*, 40 (Aug. 1991), 952-962.

[4] H-Y. LO and Y. AOKI, "Generation of a precise binary logarithm with difference grouping programmable logic array," *IEEE Trans. on Computers*, C-34 (Aug. 1985), 681-691.

[5] J. N. MITCHELL, JR., "Computer multiplication and division using binary logarithms," *IRE Trans. on Elect. Computers*, EC-11 (Aug. 1962), 512-517.

[6] E. E. SWARTZLANDER, JR., and A. G. ALEXOPOULOS, "The sign/logarithm number system," *IEEE Trans. on Computers, C-24* (Dec. 1975), 1238-1242.

[7] F. J. TAYLOR *et al.*, "A 20-bit logarithmic number system processor," *IEEE Trans. on Computers, 37* (Feb. 1988), 190-199.

[8] L. K. YU and D. M. LEWIS, "A 30-bit integrated logarithmic number system processor," *IEEE J. of Solid-State Circuits, 26* (Oct. 1991), 1433-1440.

11

THE RESIDUE
NUMBER SYSTEM

The residue number system is an integer number system whose most important property is that additions, subtractions, and multiplications are inherently carry-free. As a result we may add, subtract, or multiply numbers in one step regardless of the length of the numbers involved. Unfortunately, other arithmetic operations, like division, comparison, and sign detection, are very complex and slow. Another problem with the residue number system is that it is an integer number system and, as a result, it is very inconvenient to represent fractions. Consequently, the residue system has not been seriously considered for use in general-purpose computers. However, for some special-purpose applications such as many types of digital filters [6], in which the number of additions and multiplications is substantially greater than the number of invocations of magnitude comparison, overflow detection, division, and alike, the residue system can be very attractive.

11.1 PRELIMINARIES

A residue number system is characterized by a base that is not a single radix but an N-tuple of integers $(m_N, m_{N-1}, \cdots, m_1)$. Each of these m_i $(i = 1, 2, \cdots, N)$ is called a *modulus*. An integer X is represented in the residue number system by an N-tuple $(x_N, x_{N-1}, ..., x_1)$ where x_i is a nonnegative integer satisfying

$$X = m_i \cdot q_i + x_i \tag{11.1}$$

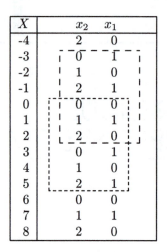

X	x_2	x_1
-4	2	0
-3	0	1
-2	1	0
-1	2	1
0	0	0
1	1	1
2	2	0
3	0	1
4	1	0
5	2	1
6	0	0
7	1	1
8	2	0

TABLE 11.1   The representation of numbers in the $(m_2, m_1) = (3, 2)$ residue system.

where q_i is the largest integer such that $0 \leq x_i \leq (m_i - 1)$. x_i is known as the residue of X modulo m_i, and the notations $X \bmod m_i$ and $|X|_{m_i}$ are commonly used.

Example 11.1

Consider a two-modulus system with the moduli $m_2 = 3$ and $m_1 = 2$. The representation of $X = 5$ in this residue system is (x_2, x_1) where

$$x_2 = |5|_3 = 2, \quad \text{since} \quad 5 = 3 \cdot 1 + 2,$$

$$x_1 = |5|_2 = 1, \quad \text{since} \quad 5 = 2 \cdot 2 + 1.$$

Therefore, the residue representation of 5 is $(2, 1)$. The number X does not necessarily have to be a positive integer. For example, if $X = -2$, then $x_1 = |-2|_2 = 0$, since $-2 = 2 \cdot (-1) + 0$. Also, $-2 = 3 \cdot (-1) + 1$, yielding $x_2 = 1$. Note that x_i is by definition positive. Thus, -2 is represented by $(x_2, x_1) = (1, 0)$.

Table 11.1 includes the representation of integers in the range $[-4, 8]$ in the $(m_2, m_1) = (3, 2)$ residue number system. As is apparent from Table 11.1, the residue representation of a number is unique. However, the converse is not true, and two or more numbers may have the same representation. For example, 1 and 7 are represented by (1,1). Consequently, we must limit the range of the numbers to be represented. As we can see from Table 11.1, the residue representation is periodic and the period in this case is 6. There are only six different representations in the residue system with the moduli $(m_2, m_1) = (3, 2)$, since x_1 can assume two possible values and x_2 can assume three. We must therefore limit the range to include only six numbers. Two such possible ranges are marked in the table. One is $-3 \leq X \leq 2$, and the other is $0 \leq X \leq 5$. □

It has been shown [7] that, in general, the number of different representations, and, as a result, the number of elements in the useful range of the residue system is the least common multiple of the moduli, denoted by

$$M = l.c.m\ (m_1, m_2, ..., m_N) \tag{11.2}$$

The least common multiple of the moduli is the smallest integer that has all the values of m_i as divisors. In the above example $M = l.c.m(2, 3) = 2 \cdot 3 = 6$, but for $m_1 = 2$ and $m_2 = 4$, $M = l.c.m(2, 4) = 4$. Hence, in order to get the largest possible range

$$M = \prod_{i=1}^{N} m_i \tag{11.3}$$

we must select moduli that are *pairwise relatively prime*. Two moduli m_i and m_j are said to be relatively prime if 1 is their greatest common divisor. This is usually written as $g.c.d(m_i, m_j) = 1$. For example, 4 and 9 are relatively prime, although neither in itself is prime.

For a given M, if only nonnegative integers are needed, the range can be set to $[0, M - 1]$. If, on the other hand, negative numbers are also desired, then the range can be set to $[-(M-1)/2, (M-1)/2]$ if M is odd, or $[-M/2, (M/2-1)]$ if M is even.

Examining the entries of Table 11.1 in the range $[0, 5]$ we should realize that a magnitude comparison between two numbers is not simple. For example, (2,1) represents a number that is larger than the number represented by (2,0), but (1,1) represents a number smaller than that represented by (0,1). This stems from the fact that, unlike the conventional number systems, the residue system is not weighted. Also, if negative numbers are included in the range then the sign of a number is not apparent from its residue representation.

11.2 ARITHMETIC OPERATIONS

The basis for performing an addition in the residue system is the identity

$$|X + Y|_{m_i} = \left|\ |X|_{m_i} + |Y|_{m_i}\ \right|_{m_i} = |x_i + y_i|_{m_i} \tag{11.4}$$

or, in general, when adding the k operands $X_1, X_2, \cdots X_k$

$$\left| \sum_{j=1}^{k} X_j \right|_{m_i} = \left| \sum_{j=1}^{k} |X_j|_{m_i} \right|_{m_i} \tag{11.5}$$

Similarly, the identity for multiplication is

$$|XY|_{m_i} = ||X|_{m_i} \cdot |Y|_{m_i}|_{m_i} = |x_i \cdot y_i|_{m_i} \tag{11.6}$$

or, in general,

$$\left| \prod_{j=1}^{k} X_j \right|_{m_i} = \left| \prod_{j=1}^{k} |X_j|_{m_i} \right|_{m_i} \tag{11.7}$$

The proofs of the above equations are straightforward [7].

Example 11.2

We add the numbers $X = 1$ and $Y = 2$ in the $(m_2, m_1) = (3, 2)$ residue system. The representations for X and Y are $(1, 1)$ and $(2, 0)$, respectively. Therefore,

$$|x_2 + y_2|_{m_2} = |1 + 2|_3 = 0$$
$$|x_1 + y_1|_{m_1} = |1 + 0|_2 = 1$$

The final result is thus $(0,1)$, which represents the value 3. Multiplying the two numbers X and Y yields

$$\left. \begin{array}{l} |x_2 \cdot y_2|_{m_2} = |1 \cdot 2|_3 = 2 \\ |x_1 \cdot y_1|_{m_1} = |1 \cdot 0|_2 = 0 \end{array} \right\} \implies X \cdot Y = (2, 0) \text{ representing the value 2.}$$

□

For subtraction we define the *additive inverse* of a number c modulo m_i as follows:

$$|-c|_{m_i} = |m_i - c|_{m_i} \qquad (\text{since } |m_i|_{m_i} = 0) \tag{11.8}$$

For example, $|-2|_3 = |3 - 2|_3 = 1$. In other words, the inverse of a number may be formed by "complementing" each residue with respect to its modulus. As for addition, the equation for subtraction is

$$|X - Y|_{m_i} = ||X|_{m_i} - |Y|_{m_i}|_{m_i} = |x_i - y_i|_{m_i} \tag{11.9}$$

Using the definition of the additive inverse, the term $|x_i - y_i|_{m_i}$ can be replaced by $|x_i + |-y_i|_{m_i}|_{m_i}$. For example, subtracting $Y = 3$ from $X = 5$ in the $(m_2, m_1) = (3, 2)$ residue system yields

$$\left. \begin{array}{l} |x_2 - y_2|_{m_2} = |2 - 0|_3 = |2 + 0|_3 = 2 \\ |x_1 - y_1|_{m_1} = |1 - 1|_2 = |1 + 1|_2 = 0 \end{array} \right\} \implies X - Y = (2, 0), \text{ representing the value 2.}$$

Example 11.3

Consider the residue number system with the set of four moduli $(m_4, m_3, m_2, m_1) = (7, 5, 3, 2)$. These moduli are pairwise relatively prime, and therefore

$$M = l.c.m(m_1, \cdots, m_4) = \prod_{i=1}^{4} m_i = 210.$$

Performing the addition and multiplication of the two operands $X = 3$ and $Y = 4$, represented by (3,3,0,1) and (4,4,1,0), respectively, yields

$$\begin{array}{ccc}
 & (7\ 5\ 3\ 2) \\
3 & (3\ 3\ 0\ 1) \\
4 \quad + & (4\ 4\ 1\ 0) \\
\hline
7 & (0\ 2\ 1\ 1)
\end{array}
\qquad
\begin{array}{ccc}
 & (7\ 5\ 3\ 2) \\
3 & (3\ 3\ 0\ 1) \\
4 \quad \times & (4\ 4\ 1\ 0) \\
\hline
12 & (5\ 2\ 0\ 0)
\end{array}$$

One can verify that the results (0,2,1,1) and (5,2,0,0) represent the expected values 7 and 12, respectively. But, when the following addition is performed,

$$\begin{array}{cc}
 & (7\ 5\ 3\ 2) \\
206 & (3\ 1\ 2\ 0) \\
7 \quad + & (0\ 2\ 1\ 1) \\
\hline
\text{should be } 213 & (3\ 3\ 0\ 1)
\end{array}$$

the result (3,3,0,1) represents the value 3, which satisfies $3 = |213|_{210}$ and we clearly have an overflow situation, which is difficult to identify.   □

11.2.1 Multiplicative Inverse

The multiplicative inverse of a number c modulo m is a number $b, 0 \leq b \leq (m-1)$ satisfying $|cb|_m = 1$, b is denoted by $|\frac{1}{c}|_m$. Any number c has an additive inverse $|-c|_m$, but the multiplicative inverse $|\frac{1}{c}|_m$ does not always exist.

The inverse $|\frac{1}{c}|_m$ exists if and only if $g.c.d(c, m) = 1$ and $|c|_m \neq 0$. If these conditions are satisfied then $|\frac{1}{c}|_m$ is unique. For example,

$m = 5$		$m = 6$		
c	$\left|\frac{1}{c}\right|_m$	c	$\left|\frac{1}{c}\right|_m$	
1	1	1	1	
2	3	2	None	$g.c.d(2, 6) = 2$
3	2	3	None	$g.c.d(3, 6) = 3$
4	4	4	None	$g.c.d(4, 6) = 2$
		5	5	

If m is a prime number, then for every possible value c satisfying $1 \leq c \leq m - 1$, $g.c.d(c, m) = 1$ and the multiplicative inverse exists.

11.3 THE ASSOCIATED MIXED-RADIX SYSTEM

Magnitude comparison, sign detection, and overflow detection for the residue
number system can be facilitated by converting the given residue representations
into the associated mixed-radix number system. This is a weighted number
system, with the representation for a number X given by

$$X = a_N \cdot (m_{N-1} \cdot m_{N-2} \cdots m_1) + \cdots + a_3 \cdot (m_2 \cdot m_1) + a_2 \cdot m_1 + a_1 \quad (11.10)$$

with the digits a_i satisfying

$$0 \leq a_i < m_i; \qquad i = 1, 2, \cdots, N. \tag{11.11}$$

Being a weighted number system implies that magnitude comparison is straightforward. For example, the values 0,1,2,3,4 and 5 in the mixed-radix system associated with the (3,2) residue system (see Table 11.1) are represented by (0,0),
(0,1), (1,0), (1,1), (2,0), and (2,1), respectively. The value of a pair (a_2, a_1) in
this mixed-radix system is $2 \cdot a_2 + a_1$.

Example 11.4

In the mixed-radix system associated with the $(m_4, m_3, m_2, m_1) = (7, 5, 3, 2)$
residue system, a number X is represented by (a_4, a_3, a_2, a_1), where

$$X = 30 \cdot a_4 + 6 \cdot a_3 + 2 \cdot a_2 + a_1$$

and the digits a_i satisfy $0 \leq a_4 < 7$, $0 \leq a_3 < 5$, $0 \leq a_2 < 3$ and $0 \leq a_1 < 2$.
The numbers 43 and 37 are represented by (1,3,1,1) and (2,2,1,1) in the
given residue system, respectively. The corresponding representations in
the associated mixed-radix system are (1,2,0,1) and (1,1,0,1), respectively.
These last two representations can be compared indicating that 43 is
greater than 37.   □

Any two numbers in a given residue system can be compared by converting
them into the associated mixed-radix system. Converting a number represented
by $(x_N, x_{N-1}, \cdots, x_1)$ in the residue system to the associated mixed-radix representation $(a_N, a_{N-1}, \cdots, a_1)$ is performed using the following equations [7]:

$$
\begin{aligned}
a_1 &= X \bmod m_1 = x_1 \\
a_2 &= (X - a_1)\left|\frac{1}{m_1}\right| \bmod m_2 \\
a_3 &= \left((X - a_1)\left|\frac{1}{m_1}\right| - a_2\right)\left|\frac{1}{m_2}\right| \bmod m_3 \\
&\vdots
\end{aligned}
\tag{11.12}
$$

This calculation can be done in residue arithmetic, as can be easily verified through the following representation of the procedure in Equation (11.12):

$$Y_{i+1} = (Y_i - a_i)\left|\frac{1}{m_i}\right| \quad \text{with} \ Y_1 = X$$

$$a_i = Y_i \bmod m_i \tag{11.13}$$

Example 11.5

To convert a number X represented by (x_4, x_3, x_2, x_1) in the residue system with the moduli $(m_4, m_3, m_2, m_1) = (7, 5, 3, 2)$ to the associated mixed-radix system, the following equations can be used:

$$a_1 = X \bmod 2 = x_1,$$

$$a_2 = (X - a_1)\left|\frac{1}{2}\right| \bmod 3,$$

$$a_3 = \left((X - a_1)\left|\frac{1}{2}\right| - a_2\right)\left|\frac{1}{3}\right| \bmod 5,$$

$$a_4 = \left(\left((X - a_1)\left|\frac{1}{2}\right| - a_2\right)\left|\frac{1}{3}\right| - a_3\right)\left|\frac{1}{5}\right| \bmod 7.$$

It is more convenient to follow the algorithm in Equation (11.13) and execute the conversion in the residue system. For example, we convert the number 43 represented by (1,3,1,1) as follows:

$$Y_1 = (1, 3, 1, 1) \quad \text{and therefore,} \ a_1 = Y_1 \bmod 2 = x_1 = 1.$$

To obtain Y_2 we first subtract a_1 from Y_1, yielding $(0, 2, 0, -)$. Note that only the first three digits in Y_2 are of interest, since a_1 is already known. We then multiply by $\left|\frac{1}{2}\right|$, which equals $(4, 3, 2, -)$, obtaining $Y_2 = (0, 1, 0, -)$. Thus, $a_2 = Y_2 \bmod 3 = 0$. Subtracting $a_2 = 0$ yields $(0, 1, -, -)$. Next we multiply by $\left|\frac{1}{3}\right|$, which equals $(5, 2, -, -)$, yielding $Y_3 = (0, 2, -, -)$. Therefore, $a_3 = Y_3 \bmod 5 = 2$. Subtracting $a_3 = 2$ we get $(5, -, -, -)$. We then multiply by $\left|\frac{1}{5}\right|_7 = 3$, yielding $Y_4 = (1, -, -, -)$. Thus, $a_4 = 1$ and the representation of 43 in the mixed-radix system is $(a_4, a_3, a_2, a_1) = (1, 2, 0, 1)$. □

The mixed-radix system is useful for overflow detection as well. For this purpose, we should add a redundant modulus m_{N+1} to the basic set of N moduli. Here, the term *redundant modulus* means that we use only the range determined by the original N moduli. For overflow detection we convert the given representation $(x_{N+1}, x_N, \cdots, x_1)$ to the associated mixed-radix system. If $a_{N+1} \neq 0$ then an overflow has occurred.

11.4 CONVERSION OF NUMBERS FROM/TO THE RESIDUE SYSTEM

If the moduli m_i are pairwise relatively prime we can use the Chinese Remainder Theorem in order to convert a number in the residue system to the conventional number system. This theorem states that

$$|X|_M = \left| \sum_{j=1}^{N} \hat{m}_j \left| \frac{x_j}{\hat{m}_j} \right|_{m_j} \right|_M \tag{11.14}$$

where $\hat{m}_j = \frac{M}{m_j}$, $M = \prod_{j=1}^{N} m_j$ and all the values of m_j are pairwise relatively prime. The proof is found in [7].

Example 11.6

For the residue number system with the three pairwise relatively prime moduli $(m_3, m_2, m_1) = (7, 3, 2)$, the range includes $M = 42$ numbers. Given a representation $(x_3, x_2, x_1) = (0, 2, 1)$, we wish to find X :

$$\hat{m}_1 = \frac{M}{m_1} = \frac{42}{2} = 21; \quad \hat{m}_2 = \frac{42}{3} = 14; \quad \hat{m}_3 = \frac{42}{7} = 6$$

$$\left| \frac{1}{\hat{m}_1} \right|_{m_1} = \left| \frac{1}{21} \right|_2 = 1; \quad \left| \frac{1}{\hat{m}_2} \right|_{m_2} = \left| \frac{1}{14} \right|_3 = 2; \quad \left| \frac{1}{\hat{m}_3} \right|_{m_3} = \left| \frac{1}{6} \right|_7 = 6$$

Therefore, $|X|_{42} = |36 \cdot x_3 + 28 \cdot x_2 + 21 \cdot x_1|_{42} = |36 \cdot 0 + 28 \cdot 2 + 21 \cdot 1|_{42} = |77|_{42} = 35$. The coefficients 36, 28, and 21 are constants that can be computed once and stored.   □

An alternate form for the equation for converting a number in the residue system to decimal is

$$|X|_M = |A_3 x_3 + A_2 x_2 + A_1 x_1|_M \tag{11.15}$$

where A_i is the "weight" of the digit x_i. Therefore, A_3 is the value of $(1, 0, 0)$, A_2 is the value of $(0, 1, 0)$ and A_1 is the value of $(0, 0, 1)$. For the residue system $(m_3, m_2, m_1) = (7, 3, 2)$ these values are 36, 28, and 21, respectively, yielding the exact same expression as in the previous example.

11.4.1 Conversion from Binary to the Residue System

If the operands are given in the conventional binary system, we can convert them directly into the residue system. Given $X = \sum_{j=0}^{n} x_j 2^j$ with $x_j \in \{0, 1\}$, then

$$|X|_m = \left| \sum_{j=0}^{n} x_j \left| 2^j \right|_m \right|_m \tag{11.16}$$

The terms $\left| 2^j \right|_m$ can be precalculated and stored in a table.

Example 11.7

To find $|1101101|_3$, we first generate a table of $|2^j|_3$, yielding

$$\left| 2^0 \right|_3 = 1, \quad \left| 2^1 \right|_3 = 2, \quad \left| 2^2 \right|_3 = 1, \quad \left| 2^3 \right|_3 = 2, \quad \cdots$$

Therefore, $|1101101|_3 = \left| 2^6 + 2^5 + 2^3 + 2^2 + 2^0 \right|_3 = |1+2+2+1+1|_3 = 1.$ □

11.5 SELECTING THE MODULI

We may have different objectives when selecting the moduli. If our objective is to reduce the execution time of addition and multiplication, then a large number of small moduli is desirable, since the execution time of these operations is determined by the largest modulus. However, a large number of small moduli will lengthen the time for converting residue numbers to the associated mixed-radix system, since this conversion is a sequential procedure in which the number of steps is determined by the number of moduli. Such conversions are necessary for magnitude comparison, sign detection, or overflow detection.

Another consideration when selecting moduli is the fact that the residues would normally be coded in some binary code, and the arithmetic operations on the residues would be executed on their corresponding binary representations. We therefore have the following objectives:

1. Efficient binary representation to minimize the total number of bits.

2. Convenient binary coding to simplify the execution of arithmetic operations.

The smallest number of bits needed to represent the residue digit for the modulus m_i is $\lceil \log_2 m_i \rceil$. Hence, to maximize the representation (storage) efficiency, we prefer to select an m_i that equals 2^k or is very close to it; e.g., $(2^k - 1)$. Clearly, we can select only one m_i of the form 2^k and still have relatively prime moduli. We may then, in addition to 2^k, select $(2^k - 1)$ and a few other moduli of the form $(2^l - 1)$. However, not all terms of the form $(2^l - 1)$ may be selected, since $2^k - 1 = (2^{k/2} - 1)(2^{k/2} + 1)$ for even values of k. Thus, $(2^k - 1)$ and $(2^{k/2} - 1)$ are not relatively prime. $(2^k - 1)$ is also factorable for some odd values of k. The selected moduli should be as close as possible to one another to avoid very large moduli, which would increase the execution time.

Example 11.8

Consider the four moduli $32 = 2^5$, $31 = (2^5 - 1)$, $15 = (2^4 - 1)$, and $7 = (2^3 - 1)$. The total number of bits required for their representation is $5 + 5 + 4 + 3 = 17$ bits. These four moduli are relatively prime, and thus,
$$M = 2^5(2^5 - 1)(2^4 - 1)(2^3 - 1) = 2^{17} - 2^{14} - 2^{13} - \cdots > 2^{16}$$
Any binary coding of 2^{16} numbers requires at least 16 bits. Therefore, these four moduli yield a very efficient coding.   □

For moduli of the form 2^k, an ordinary binary adder can be used, in which case the additive inverse is simply the two's complement. For $(2^k - 1)$, an adder with end-around carry can be used, and the additive inverse is the one's complement.

Example 11.9

If $m = 2^l - 1$, the additive inverse of the digit c is $m - c = (2^l - 1) - c$, which equals the one's complement of c. Suppose $l = 3$ and, as a result, the modulus is 7. Also assume the conventional binary coding for the residue digits. If we wish to subtract 4 from 6, we instead add the one's complement of 4 to 6, yielding

```
        110   6
    +   011   one's complement of 100=4₁₀
    ─────────
    1   001
    ─────────
        1   End-around carry
    ─────────
        010
```
                                                                        □

For moduli different from 2^k or $(2^k - 1)$, look-up tables must be used. For example, the addition and multiplication tables for $m = 5$ are depicted in Table 11.2. Each of these tables is of size $2^6 \times 3 = 64 \times 3$ bits.

+	0	1	2	3	4
0	0	1	2	3	4
1	1	2	3	4	0
2	2	3	4	0	1
3	3	4	0	1	2
4	4	0	1	2	3

×	0	1	2	3	4
0	0	0	0	0	0
1	0	1	2	3	4
2	0	2	4	1	3
3	0	3	1	4	2
4	0	4	3	2	1

TABLE 11.2   Modulo 5 addition and multiplication tables.

Digit	0	1	2	3	4
Binary Code	000 or 111	001	010	101	110

TABLE 11.3   Alternate binary coding for residues modulo 5.

In most cases, the conventional binary coding for the digits is used. This is not really necessary and, for $m = 5$, for example, we may select the coding shown in Table 11.3. The pairs 1 and 4, and 2 and 3, are additive inverse pairs and also one's complements.

11.6   ERROR DETECTION AND CORRECTION

Two subjects are discussed in this section. The first is the use of residue arithmetic to detect and possibly correct errors when performing arithmetic operations on numbers represented in the conventional number systems. The second is the use of redundant moduli in a residue system to allow detection and possibly correction of errors while performing arithmetic operations in the residue system.

11.6.1   Error Codes for Conventional Number Systems

Arithmetic error codes are those codes that are preserved under arithmetic operations. This property enables the detection of errors immediately after the completion of the arithmetic operation. Such concurrent error detection can always be attained by duplicating the arithmetic processor. This method, however, is too costly.

We say that an error code is preserved under an arithmetic operation $\star$ if for any two operands X and Y, and the corresponding encoded entities X' and Y', there is an operation $\circledast$ for the coded operands satisfying

$$(X' \circledast Y') = (X \star Y)' \tag{11.17}$$

Error codes to be used in an arithmetic unit should be examined for implementation costs and effectiveness. By costs we mean both hardware cost and execution time cost (the additional delay due to the need to encode the operands and check the result). By effectiveness we mean fault coverage, which is defined as the percentage of possible faults that will be detected (weighted percentage considering the probability of the different faults). Single-bit faults clearly have a higher probability than multiple-bit faults, and we would like to make sure that all of them are detected by the checking scheme. Note, however, that a single error in an operand or an intermediate result may cause a multiple-digit error in the final result. For example, when adding two binary numbers, if stage i of the adder is faulty, all the remaining $(n - i)$ higher order digits may be erroneous.

There are two classes of arithmetic codes: the separate codes and the nonseparate codes. In the separate codes the data and check bits are completely separated allowing us to use the data bits immediately with no encoding. We start with the simplest nonseparate codes which are the AN-codes [1]. These codes are formed by multiplying the operands by a constant A. In other words, X' in Equation (11.17) is $A \cdot X$, and the operations $\circledast$ and $\star$ are identical. For example, if $A = 3$ we multiply each operand by 3 (obtained as $2X + X$) and check the result of any arithmetic operation to see whether it is an integer multiple of 3. All error magnitudes that are multiples of A are undetectable. Therefore, we should not select a value of A that is a power of the radix 2. An odd value of A will detect every single digit fault, since such an error has a magnitude of 2^i. $A = 3$ provides the least expensive AN code that still enables the detection of all single errors.

Example 11.10

The number $0110_2 = 6_{10}$ is represented in the AN code with $A = 3$ by $010010_2 = 18_{10}$. A fault in bit position 2^3 may result in the erroneous number $011010_2 = 26_{10}$. This error is easily detectable, since 26 is not a multiple of 3.   □

The simplest separate codes are the residue code and the inverse residue code. In each of these we attach a separate check symbol $C(X)$ to every operand X. For the residue code, $C(X) = X \bmod A$, where A is called the check modulus. For the inverse residue code, $C(X) = A - (X \bmod A)$. For both separate codes, Equation (11.17) is replaced by

$$C(X) \circledast C(Y) \;=\; C(X \star Y). \tag{11.18}$$

This equation clearly holds for addition, multiplication, and subtraction (see Equations (11.4), (11.6), and (11.9), respectively). For division, the equation $X - S = Q \cdot D$ is satisfied where X is the dividend, D is the divisor, Q is the quotient, and S is the remainder. The corresponding residue check is therefore

$$\big|\, |X|_A - |S|_A \,\big|_A = \big|\,|Q|_A \cdot |D|_A\,\big|_A \,.$$

For example, if $A = 3$, $X = 7$ and $D = 5$, the results are $Q = 1$ and $S = 2$. The corresponding residue check is: $\big|\,|7|_3 - |2|_3\,\big|_3 = \big|\,|5|_3 \cdot |1|_3\,\big|_3 = 2$.

A residue code with A as a check modulus has the same exact undetectable error magnitudes as the corresponding AN code. For example, if $A = 3$, only errors that modify the result by some multiple of 3 will go undetected, and consequently, single-bit errors are always detectable. In addition, the checking algorithms for the AN code and the residue code are the same; in both we have to compute the residue of the result modulo A. Even the increase in word length,

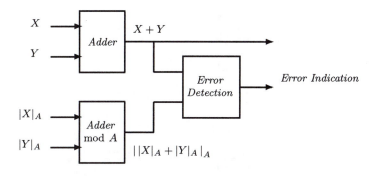

FIGURE 11.1   An adder with a separate residue check.

$|\log_2 A|$, is the same for both codes. The most important difference is due to the property of separateness. The arithmetic unit for the check symbol $C(X)$ in the residue code is completely separate from the main unit operating on X, while only a single unit (of a higher complexity) exists in the case of the AN code. An adder with a residue code is depicted in Figure 11.1. In the error detection block shown in this figure the residue modulo A of the $X+Y$ input is calculated and compared to the other input. A mismatch results in an error indication.

The AN and residue codes with $A=3$ are the simplest examples of a class of arithmetic (separate and nonseparate) codes which use a value of A of the form $A = 2^a - 1$, with a being an integer [1]. This choice simplifies the calculation of the remainder when dividing by A (which is needed for the checking algorithm), and it is the reason that these codes are called *low-cost* arithmetic codes. The calculation of the remainder when dividing by $2^a - 1$ is simple, because the equation

$$\left| z_i r^i \right|_{r-1} = \left| z_i \right|_{r-1}, \qquad r = 2^a. \tag{11.19}$$

allows the use of modulo $2^a - 1$ summation of the groups of size a bits that compose the number (each group has a value $0 \le z_i \le 2^a - 1$).

Example 11.11

To calculate the remainder when dividing the number $X = 11110101011$ by $A = 7 = 2^3 - 1$, we partition X into groups of size 3, starting with the least significant bit. This yields $X = (z_3, z_2, z_1, z_0) = (11, 110, 101, 011)$. We then add these groups modulo 7; i.e., we "cast out" 7's and add the end-around-carry whenever necessary. A carry-out has a weight of 8 and since $|8|_7 = 1$ we must add an end-around-carry whenever there is a carry-out as illustrated below.

$$
\begin{array}{rl}
11 & z_3 \\
+ \quad 110 & z_2 \\
\hline
1 \quad 001 & \\
+ \qquad 1 & \text{end-around carry} \\
\hline
010 & \\
+ \quad 101 & z_1 \\
\hline
111 & \\
+ \quad 011 & z_0 \\
\hline
1 \quad 010 & \\
\quad\ 1 & \text{end-around carry} \\
\hline
+ \quad 011 & \\
\end{array}
$$

The residue modulo 7 of X is 3, which is the correct remainder of $X = 1963_{10}$ when divided by 7. □

Both separate and nonseparate codes are preserved when we perform arithmetic operations on unsigned operands. If we wish to include signed operands as well, we must require that the code be complementable with respect to R, where R is either 2^n or $2^n - 1$ (where n is the number of bits in the encoded operand). The selected R will determine whether two's complement or one's complement arithmetic will be employed. The original operand is complementable with respect to M (as before, M is either 2^m or $2^m - 1$, but with $m < n$). Hence, for the AN code, the equation $R - AX = A(M - X)$ must be satisfied, yielding $R = AM$; i.e., A must be a factor of R. If we insist on A being odd, it excludes the choice $R = 2^n$. Thus, only one's complement can be used, with A being a factor of $2^n - 1$.

Example 11.12

For $n = 4$, R is equal to $2^n - 1 = 15$ for one's complement, and is divisible by A for the AN code with $A = 3$. The number $X = 0110$ is represented by $3X = 010010$, and its one's complement 101101 $(= 45_{10})$ is divisible by 3. However, the two's complement of $3X$ is 101110 $(= 46_{10})$ and is not divisible by 3. If $n = 5$, then for one's complement R is equal to 31, which is not divisible by A. The number $X = 00110$ is represented by $3X = 0010010$, and its one's complement is 1101101 $(= 109_{10})$, which is not divisible by 3. □

For the residue code with the check modulus A, the equation $A - C(X) = (R - X) \bmod A$ must be satisfied. This implies that R must be an integer multiple of A, again allowing only one's complement arithmetic to be used. However, we may modify the procedure so that two's complement (with $R = 2^n$) can also be employed:

$$(2^n - X) \bmod A = (2^n - 1 - X + 1) \bmod A = (2^n - 1 - X) \bmod A + 1 \bmod A$$
$$(11.20)$$

We therefore, need to add a correction term $|1|_A$ to the residue code when forming the two's complement. Note that A must still be a factor of $2^n - 1$. A similar correction is needed when we add operands represented in two's complement and a carry-out (of weight 2^n) is generated in the main adder. Such a carry-out is discarded according to the rules of two's complement arithmetic. To compensate for this, we need to subtract $|2^n|_A$ from the residue check. Since A is a factor of $(2^n - 1)$, the term $|2^n|_A$ is equal to $|1|_A$.

These modifications result in an interdependence between the main arithmetic unit and the check unit that operates on the residues. Such an interdependence may cause a situation where an error from the main unit propagates to the check unit and the effect of the fault is masked. However, it has been proven in [2] that the occurrence of a single-bit error is always detectable.

Example 11.13

For the residue code with $A = 7$ and $n = 6$, $R = 2^6 = 64$ for two's complement and $R - 1 = 63$ is divisible by 7. The number $001010_2 = 10_{10}$ has the residue 3 modulo 7. The two's complement of 001010 is 110110. The complement of $|3|_7$ is $|4|_7$ and adding the correction term $|1|_7$ yields 5, which is the correct residue modulo 7 of 110110 ($= 54_{10}$).

If we now add to $X = 110110$ (in two's complement) the number $Y = 001101$, a carry-out is generated and discarded. We must therefore subtract the correction term $|2^6|_7 = |1|_7$ from the residue check with the modulus $A = 7$, obtaining

$$
\begin{array}{r}
110110=X \\
+ \quad 001101=Y \\
\hline
1 \quad 000011
\end{array}
\qquad
\begin{array}{r}
101=|X|_7 \\
+ \quad 110=|Y|_7 \\
\hline
1 \quad 011 \\
\quad 1 \;\; \text{end-around carry} \\
\hline
100 \\
- \quad 1 \;\; \text{correction term} \\
\hline
011
\end{array}
$$

3 is the correct residue of the result 000011 modulo 7.   □

Error correction can be achieved by using two or more residue checks. The simplest case is the biresidue code, which consists of two residue checks A_1 and A_2. If $A_1 = 2^a - 1$ and $A_2 = 2^b - 1$ are two low-cost residue checks with $n = l.c.m.(a, b)$ where n is the number of bits in the operands, then any single-bit error can be corrected [5].

11.6.2 Error Codes for the Residue Number System

The residue system is inherently more fault-tolerant than the conventional number system. The lack of interaction among the residue digits (no carry-propagation) implies that a fault in a single digit will not result in errors in other digits. This desirable fault isolation property is preserved while performing addition, subtraction, and multiplication, but is not preserved while other operations are performed. Another consequence of the fault isolation property is that a consistently erroneous residue digit, when identified, can be disconnected and the rest of the residue arithmetic unit can still be used.

The fault-tolerance features will be manifested only if redundant moduli are added to the original set of moduli, allowing the detection of errors and even the identification of faulty residue digit circuits. The resulting system is called a *redundant residue number system* and is defined by a set of $N + L$ moduli $(m_{N+L}, \cdots, m_{N+1}, m_N, \cdots, m_1)$. The L moduli $m_{N+L}, \cdots, m_{N+1}$ are the redundant moduli, implying that out of the total range $[0, M_T - 1]$ (where $M_T = \prod_{i=1}^{N+L} m_i$) $[0, M - 1]$ (with $M = \prod_{i=1}^{N} m_i$) is the legitimate range. The range $[M, M_T - 1]$ is called the illegitimate range. It has been shown [6] that a single error always moves the operand from the legitimate range to the illegitimate range and can therefore be easily identified.

11.7 EXERCISES

11.1. Verify that there are only four different representations in the residue number system with $(m_2, m_1) = (4, 2)$.

11.2. Given the set of moduli (7,5,3), find: (a) the range M, (b) the coefficients for the Chinese Remainder Theorem and the value represented by (2,3,2), (c) the corresponding mixed-radix representation, (d) the representation of 20 in the residue system and in the mixed-radix system.

11.3. Prove the Chinese Remainder Theorem by calculating the remainder modulo m_i of Equation (11.14), knowing that every number X has a unique representation in the residue system.

11.4. Write the subtraction table for $m = 5$ first in decimal representation and then in binary representation using (a) the ordinary binary coding with 000 through 101 (for the digits 0 through 4, respectively), (b) the binary code in Table 11.3. What is the advantage of the second scheme?

11.5. Write the rule for converting a decimal number to the residue system using a table of $|10^j|_m$. How is this rule simplified for the case $m = 9$?

11.6. Divide 35 by 5 in the residue system with $(m_3, m_2, m_1) = (7, 3, 2)$. Can you divide 35 by 7? 34 by 5? What are the conditions under which division can easily be carried out?

11.7. Given a number X and its residue modulo 3, $C(X) = |X|_3$. How will the residue change when X is shifted by one bit position to the left if the shifted-out bit is 0? Repeat this for the case where the shifted-out bit is 1. Verify your rule for $X = 01101$ shifted five times to the left.

11.8. The calculation of the remainder when dividing by $A = 2^a - 1$ can be done in parallel rather than in series manner. Show a block diagram of such a parallel circuit for 32-bit long numbers and $A = 15$.

11.9. Show that a residue check with the modulus $A - 2^a - 1$ can detect all errors in a group of $a - 1$ (or less) adjacent bits. Such errors are called *burst errors* of length $a - 1$ (or less) and they may occur when shifting an operand by several bit positions.

11.10. Prove that $|z_i r^i|_{r-1} = |z_i|_{r-1}$ for $r = 2^a$ and $0 \le z_i \le 2^a - 1$.

11.11. When performing a divide operation with AN coded operands the quotient Q must be multiplied by A. Find a simple algorithm for executing this multiplication when $A = 2^a - 1$. Illustrate your algorithm for $A = 15$.

11.8 REFERENCES

[1] A. AVIZIENIS, "Arithmetic error codes: Cost and effectiveness studies for application in digital system design," *IEEE Trans. on Computers, C-20* (Nov. 1971), 1322-1331.

[2] A. AVIZIENIS, "Arithmetic algorithms for error-coded operands," *IEEE Trans. on Computers, C-22* (June 1973), 567-572.

[3] Y. MA, *IEEE Trans. on Computers, 47,* (March 1998), 333-337.

[4] F. POURBIGHARAZ, and H. M. YASSINE, "A signed-digit architecture for residue to binary transformation," *IEEE Trans. on Computers, 46* (Oct. 1997), 1146-1150.

[5] T. R. N. RAO, "Biresidue error-correcting codes for computer arithmetic," *IEEE Trans. on Computers, C-19* (May 1970), 398-402.

[6] M. A. SODERSTRAND, W. K. JENKINS, G. A. JULLIEN, and F. J. TAYLOR, *Residue number system arithmetic modern application in digital signal processing,* IEEE Press, New York, 1986.

[7] N. S. SZABO and R. I. TANAKA, *Residue arithmetic and its application to computer technology,* McGraw-Hill, New York, 1967.

[8] F. J. TAYLOR, "Residue arithmetic: A tutorial with examples," *IEEE Computer, 17* (May 1984), 50-62.

[9] R. ZIMMERMANN, "Efficient VLSI implementations of modulo $(2^n \pm 1)$ addition and multiplication," *Proc. of 14th Symp. on Computer Arithmetic* (April 1999), 158-167.

INDEX